Water in the Middle East

Springer
Berlin
Heidelberg
New York
Barcelona
Budapest
Hong Kong
London
Milan
Paris
Santa Clara
Singapore
Tokyo

Waltina Scheumann
Manuel Schiffler (Eds.)

Water in the Middle East

Potential for Conflicts and Prospects for Cooperation

International Water Convention
and water-related treaties are documented

Springer

Editors:

Dr.-Ing., Dipl. Pol. Waltina Scheumann
Technical University of Berlin, IMUP
FR 2-7, Franklinstr. 28/29
10587 Berlin
Germany

Dr. rer. pol. Manuel Schiffler
The World Bank
1818 H-Street, N.W.
Washington, D.C. 20433
USA

With 6 Figures and 8 Tables

ISBN 3-540-64029-0 Springer-Verlag Berlin Heidelberg New York

Cataloging-in-Publication Data applied for

Water in the Middle East : potential for conflicts and prospects for cooperation ; with 8 tables / Waltina Scheumann ; Manuel Schiffler (ed.). - Berlin ; Heidelberg ; New York ; Barcelona ; Budapest ; Hong Kong ; London ; Milan ; Paris ; Santa Clara ; Singapore ; Tokyo :
Springer, 1998
ISBN 3-540-64029-0

Typesetting: Camera-ready by the editors
Cover design: *de'blik, Berlin*
SPIN: 10565565 30/3136-5 4 3 2 1 0 – Printed on acid-free paper.

Acknowledgements

The editors would like to thank all the participants at the first and second conference on Water in the Middle East in Berlin, held on June 17-18, 1995 and September 21-22, 1996 respectively. Their contributions have been crucial in elaborating and testing the ideas contained in this volume, which have been updated ever since for this publication. In particular, we would like to express our gratitude to Sebastian Büttner, at that time at the Science Center Berlin (WZB), who moderated the two conferences with great diligence and undeniable charm.

Berlin 1998

Waltina Scheumann
Manuel Schiffler

Contents

Part II: Regional Studies

Abbreviations

BCM	billion cubic metres
EU	European Union
GAP	Turkish acronym for the Southeastern Anatolian Project
GDI	German Development Institute
GDP	Gross Domestic Product
GNP	Gross National Product
GTZ GmbH	German acronym for the German Society for Technical Co-operation
ICPR	International Commission for the Protection of the Rhine
ILA	International Law Association
ILC	International Law Commission
MCM	million cubic metres
MW	Megawatt
MWGWR	Multilateral Working Group on Water Resources
OAU	Organization of African Unity
PKK	Kurdish Workers' Party
PLO	Palestine Liberation Organization
UNDP	United Nations Development Programme
UNEP	United Nations Environment Programme
WHO	World Health Organization

List of Illustrations

List of Contributors

Jörg Barandat, Dipl.Päd., LTC(GS)
Federal Armed Forces -
General Staff College
Manteuffelstr. 20
22587 Hamburg
Germany

Ines Dombrowsky, Dipl.Ing.
The World Bank
1818 H-Street, N.W.
Washington, D.C. 20433
USA

Rainer Durth, Dr.rer.pol.
The German Development Bank
Palmengartentr. 5-9
60325 Frankfurt/Main
Germany

Aytül Kaplan, LL.M., Lawyer
Türkiye Halk Bankasi
Deutschland Repräsentanz
Marzellenstr.9
50667 Köln
Germany

Ulrich Küffner, Dr.-Ing. (Ph.D.),
Dipl.Ing. (M.Sc.),
Avenida Colon 1370, Dpt.15-A
Quito
Ecuador

Jochen Renger, Dipl.Pol.
Heinrichsweg 3/28
72074 Tübingen
Germany

Waltina Scheumann, Dr.-Ing., Dipl.Pol.
Technical University of Berlin, IMUP
FR 2-7, Franklinstr. 28/29
10587 Berlin
Germany

Manuel Schiffler, Dr.rer.pol.
The World Bank
1818 H-Street, N.W.
Washington, D.C. 20433
USA

Introduction

The former Egyptian Minister of State for Foreign Affairs and former UN Secretary General, Butros Butros Ghali stated after the second Gulf War „The next war in the Middle East will not be fought for oil, but for water."

This famous statement has been echoed by many politicians: shortly before becoming president of Turkey, Süleyman Demirel declared that the waters of the Euphrates and Tigris belonged to Turkey, just as oil belongs to the Arabs. Rafael Eytan, at that time and now again Israeli Minister of Agriculture, declared in 1990 in full-page advertisements in the Jerusalem Post that Israel would never cede the West Bank to the Palestinians because Israel's water supply would otherwise be endangered. Finally, Ismail Serageldin, vice president of the World Bank, declared in 1995 that „the wars of the next century will be over water".

These statements are typical of the atmosphere reigning in the Middle East and in several other places around the world concerning the issue of international fresh water resources. Whether these perceptions correspond to an actual threat to a nation's water supply or whether they correspond to the official position of states in negotiations often conducted secretly, is an entirely different matter. A closer analysis of the issue of international fresh water resources, as we attempt in this book, admittedly reveals a dangerous potential for conflict over water. However, it also places these conflicts within the larger framework of entrenched political conflicts in the region and shows past and present attempts of solutions, as well as a future potential for cooperation and possible win-win solutions for the parties concerned.

In doing so, the analysis and its conclusions are not limited to the Middle East. An estimated 40% of the world's population live in the watersheds of international rivers. International water conflicts occur in Asia as well, as shown for example by the Indus, Ganges and Mekong rivers. Other international rivers, such as the Senegal, Niger and Zambezi rivers in Africa and the Rio Grande and Paraná rivers in the Americas, have given less rise to tensions up to now, but a potential for conflict prevails there as well. In Europe, however, issues concerning international rivers, among which are mainly issues of navigation and water quality along the Rhine and the Danube, have been resolved relatively successfully.

International conferences have repeatedly focused on the issue of improved water management, including the issue of international rivers. Non-governmental professional organizations have been among the first to grasp the problem. For example, in 1966 the International Law Association adopted the Helsinki Rules, which set forth recommendations for the equitable utilization of international fresh water resources. Intergovernmental conferences followed suit. In 1977, the first

United Nations Water Conference united representatives of the world's governments in Argentina to adopt the Mar del Plata Action Plan, which later gave rise to the International Decade on Water and Sanitation. The Mar del Plata Action Plan was the first intergovernmental document to call for an equitable management of international fresh water resources. A few years earlier, in 1970, the International Law Commission of the United Nations had been charged with drafting a convention on international fresh water, a task which took more than a generation to accomplish. Meanwhile, the International Conference on Water and the Environment adopted the Dublin Statement in 1992 and in the same year the United Nations Conference on Environment and Development (UNCED) in Rio de Janeiro gave birth to Agenda 21, the global environmental action plan for the 21st century which included a chapter on the management of fresh water resources. Both documents called for the integrated management of fresh water resources at the level of the river basin, be it within a single country or internationally. Following these conferences, new international institutions incorporating international professional organizations and donor agencies as well as the private sector were set up, such as the World Water Council and the Global Water Partnership, which were both founded in 1996. The first World Water Forum, which was organized by these new institutions, convened in March 1997 in Marrakesh and reiterated many of the propositions of earlier conferences. Finally, in May 1997, after lengthy and controversial debates, the General Assembly of the United Nations adopted the International Water Convention by a large majority, setting up rules for the equitable and reasonable utilization of international fresh water resources. This can be considered a major step forward, taking into account that this is the first document showing a large intergovernmental consensus on this crucial issue. The Convention, however, only becomes effective after ratification by 35 countries, which is expected to happen in the next 3 years.

The region with the world's most acute water conflicts undoubtedly is the Middle East with its three major international river basins, the Jordan, the Euphrates and the Nile. It remains a matter of scholarly dispute whether conflicts over scarce water resources are actually a major reason for tension and armed conflict in the region or will become so. The opposite argument maintains that conflict in the Middle East is deeply entrenched for various historical and political reasons that are totally unrelated to water scarcity and that water conflicts are simply one arena in which these conflicts are carried out. Other water conflicts in arid areas that lack such political conflicts, such as the conflict between the United States and Mexico, have never led to tensions of the same magnitude. Whether natural resource conflicts cause political conflicts or 'purely' engender or exacerbate resource conflicts is nevertheless difficult to establish, and no definite answer can be expected from this volume, which tries to incorporate various views on this subject.

The approach of this volume is interdisciplinary. In the view of the authors, political scientists, economists, lawyers and engineers, any unidisciplinary approach to the extremely complex international water issues would be flawed.

The present volume is the result of an academic conference organized by the German Development Institute (GDI) and the Institute for Management in Environmental Planning (IMUP) of the Technical University of Berlin on September

21-22, 1996, at the GDI in Berlin. The conference followed an earlier conference at the GDI on June 17-18, 1995, on water management in the Jordan Basin, the results of which have already been published.[1] The intensive exchange of views during and after these two conferences enabled the authors to incorporate many suggestions from participants into their papers, so that we hope to have produced a volume that does not only group together a number of loosely related topics, but presents a coordinated view as well as, of course, different points of view advocated by the authors.

The research on this volume has, of course, not taken place in an academic vacuum. A number of books have been published during the past years on water in the Middle East. Most notable is the research undertaken in the United States in the mid-1980s by Naff and Matson, Kolars and Mitchell, as well as Starr and Stoll, who were followed by Lowi and Wolf, among others, in the early 1990s. Remarkable contributions in the 1990s have come as well from Allan, Mallat and Biswas in England, Kliot in Israel and Ohlsson and others in Sweden.[2] In Germany and Switzerland, research on this topic did not emerge until the mid-1990s,[3] and built on the experience gained by earlier researchers to whom we are indebted. This volume attempts to draw together the results of this research, while insisting on the interdisciplinary nature of the subject and incorporating the practical experience of the authors, many of whom have had first hand professional experience with water resource management in the Middle East.

This volume is divided into two parts. The first part deals with the more general legal, political and economic aspects of international water conflicts, while the second part presents regional studies on the Jordan, Euphrates and Nile basins.

The first part begins with a review of current international fresh water law by *Barandat* and *Kaplan*. They reiterate the three different fundamental principles of that law and provide an overview of its historical development since the adoption of the Helsinki Rules. Of special interest is their inside view of the heated discussions in the United Nations leading up to the International Water Convention, including the efforts by a number of countries to make the convention less specific and less binding and to delay of its coming into effect. The authors further discuss the scope of the enforcement of international law, including the proposed creation of a UN Environmental Council. They conclude by stressing that nation states have to relinquish some of their accustomed national sovereignty if international fresh water law is to become effective and that the spread of 'civil societies' in developing countries is a necessary precondition to making states more reliable partners in the international arena.

Schiffler builds on Barandat and Kaplan's view by analyzing five existing bilateral water treaties in the Middle East (between Egypt and Sudan, Israel and Jordan, Jordan and Syria, Syria and Iraq, as well as Turkey and Syria) with regard to six factors considered crucial to their effectiveness: their compatibility with inter-

---

1 Libiszewski and Schiffler 1996

2 Naff and Matson 1984; Kolars and Mitchell 1991; Starr and Stoll 1988; Lowi 1995; Wolf 1995; Allan and Mallat 1995; Biswas 1994; Kliot 1994; Ohlsson 1995.

3 Cf. Barandat 1997, Dombrowsky 1995, Libiszewski 1995, Scheumann 1993, Schiffler 1995; Vallianatos-Grapengeter 1996; Durth 1996.

national fresh water law, the importance of links with non-water issues, the consideration of other riparian's rights in bilateral agreements, the distribution of the risk of droughts among treaty parties, the consideration of ecological factors and the provisions for the enforcement of the treaties. He concludes by giving some recommendations for the elaboration of new international water treaties, which are considered an essential feature of any settlement of water disputes in the Middle East.

Renger draws links between the water issue and the general political situation in the Middle East, by building on the theoretical concepts of power, security and welfare. He believes the reason for the standstill of the peace process is that Israel as the strongest party has achieved its objectives to a substantial part: improvement of its relations with a number of Arab states, lifting of the Arab boycott, and other successes in the international arena such as the signing of the association agreement with the European Union in 1995. Concerning power and security, Israel's conflict with the Palestinians and Jordan is a genuine water conflict. The conflict between Israel and Syria/Lebanon, which both have significant water resources outside the Jordan Basin is, however, a conflict mainly concerning general security issues on the Israeli side and issues of national prestige on the Arab side. Although half of Israel's water supply depends on water resources located outside its internationally recognized borders, the possibility of fighting over water seems marginal. Concerning welfare, Israel refuses to reduce its agricultural water use. The Palestinians, however, whose per capita water use is much lower, suffer the most from the consequences of water scarcity. Politically, the Palestinians are weak and are far from achieving their main objective in the peace process, i.e. a sovereign state of Palestine. Syria has found it comfortable not to join the peace process, and given its present state, this position is unlikely to change. Renger concludes that there is some limited scope for regional cooperation in the water sector, but that questions of security and territory have precedence over questions of water.

Durth compares international water conflicts in politically and economically 'integrated' areas such as Europe to water conflicts in 'non-integrated' areas such as the Middle East. In non-integrated areas, different perceptions of what is just and equitable prevail, as opposed to a stronger consensus on that issue in integrated areas. In the former situation, an agreement on just procedures as a first step may be more feasible than an agreement on just outcomes. The position of the parties to the conflict is strongly influenced by considerations of national prestige, rather than by measurable economic benefits. These potential economic benefits of cooperation, such as Turkey's offer to sell water and electricity to Arab countries, are left unrealized. In Europe, as opposed to the Middle East, the state monopoly on information exchange and on negotiations was diluted by allowing non-governmental entities such as water utilities to enter into contact and engage in legal agreements, as was the case with the Port of Rotterdam, which concluded agreements with upstream entities responsible for the siltation of the port. The article ends with a number of recommendations for solving international water conflicts.

Küffner begins his article with an overview on international water conflicts all over the world and a brief analysis of existing international water treaties and their weaknesses. Then he advocates the concept of sharing international rivers by jointly managing them to the maximum mutual benefit through an international agency, instead of dividing the waters and managing them separately at higher costs. International experience, as exemplified by the German Ruhrverband, among others, shows that such joint management resolves disputes and leads to cost reductions. A large body of experience with such joint management approaches has been gained in the electricity sector. The experience in the water sector, however, remains limited to cooperation among water and sewage utilities at the sub-national level. The creation of international agencies grouping together national water utilities, the introduction of cost-covering tariffs, the privatization of water utilities, and the abandonment of public ownership of water resources in favour of private water rights might stimulate such a process and contribute to saving water. In the Jordan Basin, however, where public water utilities subsidize water tariffs and public ownership of water resources prevail and where trust is often absent, such a cooperation is tremendously difficult. Only a few examples of international cooperation exist and are described in the article as a possible building stone for future cooperation.

The second part of this volume contains regional studies on the Jordan, Euphrates and Nile basins.

Dombrowsky provides an overview of water resources and water use in the Jordan Basin, by drawing on the results of a recent joint Israeli, Palestinian and Jordanian regional water study financed by the German government in the framework of the multilateral peace negotiations. The regional study not only presents water data on a national basis, but also on the basis of the hydrological basin, which is widely accepted to be the appropriate geographical unit for water management. Dombrowsky compares the hydro-strategic positions of the five riparians. She goes on to describe the results of the bilateral and multilateral tracks of the peace process concerning water, including the two agreements signed, i.e. Annex II of the Treaty of Peace between Israel and Jordan and the Interim Agreement (Oslo B) between Israel and the Palestinian Authority. Dombrowsky concludes that both agreements lack respect for some principles of international water law in favour of Israel. She analyzes the potential economic benefits of cooperation, but also stresses the obstacles to cooperation in the water sector, among which are foremost the stalemate in the peace process, the reluctance to reduce agricultural water use, and economic inequalities. In conclusion, she warns that the international community should not focus on financing 'grand solutions' in infrastructure, as favoured by the partners in the region, but rather on supporting less expensive, but politically more difficult improved joint management of existing resources.

Scheumann describes the ongoing Turkish irrigation and hydropower projects on the Upper Euphrates and assesses their potential impact on the river's water balance. Surprisingly, she finds that due to the slow implementation of the irrigation schemes the current impact on the water balance is very low, and this situation will change only very slowly. Furthermore, the water claims of the riparians are not too far apart, so that an agreement seems possible if other factors do not interfere. In her analysis of the history of conflict and negotiation among the three riparians, i.e. Turkey, Syria and Iraq, over the past half century, Scheumann finds that non-water issues, such as Syrian support for the Kurdish Workers' Party PKK, strongly influenced attitudes on the water issue in a most negative way. However, at least in one case they favoured an agreement on water issues among Turkey and Syria, which shows that the overall political relations between the riparian states are crucial to cooperative results. Disputes over territorial sovereignty continue to be an impediment to an agreement on water issues, but an agreement seems possible if it provides benefits for all parties involved. As Scheumann shows, upstream development projects, despite their undisputed negative effects, bring some often overlooked benefits to downstream riparians, such as the trapping of sediments and the smoothing of fluctuations in river flow.

After a brief historical overview of water use patterns in the Nile Basin since the turn of the century, *Schiffler* analyzes the Nile Waters Agreement of 1959 between Egypt and Sudan, which remains until today the only international water sharing agreement between the ten riparians of the Nile. The crucial position of Ethiopia as the most important upstream country and the implications of the building of dams along the Blue Nile are analyzed. Furthermore, the prospects for a multilateral agreement among all the riparians as a result of an ongoing conference series are assessed. Possible impacts of climate change on the water balance are discussed as well, along with the potential effects of flow reductions on Egypt as the most donwstream riparian. Egypt's recent moves in agricultural policy, including the planned diversion of water into a 'Second Nile' in the western desert and the transfer of Nile water to the Sinai, are also critically reviewed. The article ends with a brief discussion about the degree to which the conflicts on the Nile are the result of water scarcity or rather of non-water factors.

Although the views of the contributors to this volume differ on some issues, in the editors' view the following conclusions can be derived from this book:

- The recently adopted International Water Convention of the United Nations, although not undisputed among the states of the region, provides a comprehensive and promising framework for the resolution of international fresh water conflicts in the Middle East.
- Water is scarce in the Middle East, but water scarcity is a relative concept. The demand for water is determined by its price, which is low throughout the region, and by agricultural policies, which aim for food self-sufficiency in many Arab countries. These policies exacerbate water demand far above the level necessary to meet municipal and industrial water needs which can be satisfied from existing conventional water resources in the Middle East even if the population significantly increases.
- Grand high-tech solutions of the water problem are bound to remain illusive, mainly because of their persisting high cost (e.g. for seawater desalination and long-distance transfers) or because of political conflicts involving international water transfers.
- The Middle East is characterized by a number of entrenched political and historical conflicts that are unrelated to water issues, but have serious negative consequences for the management of international water resources. However, genuine 'water wars', as have been predicted by many observers, are therefore unlikely to occur.

- An important domestic non-water issue is the influence of agricultural lobbies which oppose any reduction of agricultural water use or a cut-back in plans to increase the irrigated area and thus can form an obstacle to cooperation over international rivers.
- International non-water issues are important, but their impact on the readiness to cooperate is ambivalent. On the one hand, the desire to be viewed as a reliable partner in the international arena can be an incentive to abide by international law. On the other hand, however, security concerns or the attempt to maintain national food self-sufficiency can hamper cooperation.
- Water stirs emotions in the Middle East. Therefore, the water issue can be easily politicized and its relevance can be exaggerated for political purposes. A sober analysis of the economic costs and benefits of various options by Middle Eastern politicians is the exception rather than the rule.
- The European experience in joint water management, as illuminating as it is, can only partially be transferred to the Middle East. The prevalence of deeply rooted political conflicts, the absence of democratically legitimated rule and of civil societies in most countries of the region, and the importance of irrigated agriculture with its specific problems are the most important factors that make the application of the European experience to the region so difficult.
- Multilateral cooperation between all riparian states in a hydrological basin in the form of joint management of a river's resources, as recommended from an economic and environmental point of view, is not feasible under such circumstances if it entails a significant encroachment on national sovereignty and unless the underlying political conflicts in the Middle East are resolved beforehand.
- Successful bilateral cooperation can, however, be achieved even under such unfavourable conditions as those prevailing in the Middle East, as a number of examples have shown. Such bilateral cooperation can and should be encouraged until more favourable conditions allow a comprehensive approach.

References

Allan, J. Anthony; Mallat, Chibli (eds.) 1995: Water in the Middle East: Legal, political and commercial implications. London.

Barandat, Jörg (ed.) 1997: Wasser - Konfrontation oder Kooperation. Ökologische Aspekte von Sicherheit am Beispiel eines weltweit begehrten Rohstoffs. Nomos Baden-Baden.

Biswas, Asit K. (ed.) 1994: International Waters of the Middle East: From Euphrates-Tigris to Nile. Water Resources Management Series No.2, Oxford.

Dombrowsky, Ines 1995: Wasserprobleme im Jordanbecken: Perspektiven einer gerechten und nachhaltigen Nutzung internationaler Ressourcen. Frankfurt/Main, Bern.

Durth, Rainer 1996: Grenzüberschreitende Umweltprobleme und regionale Integration. Zur Politischen Ökonomie von Oberlauf-Unterlauf Problemen an internationalen Flüssen. Nomos Baden-Baden.

Kliot, Nuri 1994: Water Resources and Conflict in the Middle East. London, New York.

Kolars, John F.; Mitchell, William A. 1991: The Euphrates River and the Southeast Anatolia Development Project. Carbondale and Edwardsville, Ill., Southern Illinois University Press.

Libiszewski, Stephan 1995: Das Wasser im Nahostfriedensprozeß. Konfliktstrukturen und bisherige Vertragswerke unter wasserpolitischer Perspektive, in: Orient, No.4.
Libiszewski, Stephan; Schiffler, Manuel (eds.) 1996: Wasserkonflikte und Wassermanagement im Jordanbecken. Beiträge eines Colloqiums in Berlin, Juni 1995. Berlin.
Lowi, Miriam 1995: Water and Power: The politics of a scarce resource in the Jordan River basin. Cambridge.
Naff, Thomas; Matson, Ruth C. 1984: Water in the Middle East: Conflict or Cooperation. Boulder, Col., Westview Press.
Ohlsson, Leif (ed.) 1995: Hydropolitics: Conflict over water as a development constraint. London.
Scheumann, Waltina 1993: New Irrigation Schemes in Southeast Anatolia and Northern Syria: More Competition and Conflict over the Euphrates?, in: Quarterly Journal of International Agriculture, No.3.
Schiffler, Manuel 1995: Das Wasser im Nahostfriedensprozeß. Ansätze zu einer gerechten Aufteilung und Möglichkeiten zur Entschärfung der Wasserknappheit, in: Orient, No.4.
Starr, Joyce R.; Stoll, Daniel C. (eds.) 1988: The politics of scarcity: Water in the Middle East. Boulder, Colo., Westview Press.
Vallianatos-Grapengeter, Ina 1996: Die wasserstrategischen Aspekte im Sechs-Tage-Krieg und ihre Relevanz für die israelisch-arabischen Friedensgespräche, in: Libiszewski; Schiffler (eds.) 1996: Wasserkonflikte und Wassermanagement im Jordanbecken. Beiträge eines Colloquium in Berlin, Juni 1995. Berlin.
Waterbury, J. 1979: Hydropolitics of the Nile Valley. Syracuse, N.Y., Syracuse University Press.
Wolf, Aaron T. 1995: Hydropolitics along the Jordan River: The impact of scarce water resources on the Arab-Israeli conflict. Tokyo, United Nations University Press.

Part I
Legal, Political and Economic Perspectives on Water Conflicts

1 International Water Law: Regulations for Cooperation and the Discussion of the International Water Convention[1]

Jörg Barandat, Aytül Kaplan

Over the next decade, water issues in the region's three major river basins - the Jordan, the Nile, and the Tigris-Euphrates - will foster either an unprecedented degree of cooperation or a combustible level of conflict.
Sandra Postel 1992, p.74, Worldwatch Institute, Washington, D.C.

1.1 Water as a Security Issue

On 21 May 1997, the United Nations General Assembly voted to adopt the text of an international water convention that was presented by the Sixth (Legal) Committee of the General Assembly. The result of that vote was as follows:

For	103 nations	
Against	3 nations	Burundi, China, Turkey
Abstentions	27 nations	e.g. Egypt, Ethiopia, France, India, Israel, Pakistan, Uzbekistan

The current convention is awaiting final approval and was opened to signature. Thirty-five nations must agree to its terms by 20 May 2000 in order to enter the convention into force.

This convention does not represent an end solution. Conflicts rooted in the right to water sources that cross over political borders have not been successfully resolved by the convention text. The convention itself should be followed up with

[1] We would like to thank Ms Telse Kantzmeier and Ms Patricia Abild, German Maritime Headquarters, Glücksburg for the translation of this contribution, Ms Waltina Scheumann and Mr Manuel Schiffler for the critical review of the first draft, and LCDR Vincent J. van Joolen, US Navy, for his assistance in editing the final draft.

political initiatives in order for it to become effective and to guarantee basic standards of international water law now and in the future.

The following essay presents the historical development of this convention, and suggests current and future actions that must be accomplished in order to make this convention a useful instrument that will address the world's water problems in its proper context.

During the Cold War the world enjoyed a unique degree of political stability. Due to the possibility of mutual self-destruction, both superpowers had a vital interest in countering existing or newly arising conflicts. They accomplished this by organizing themselves into the bipolar East-West pattern, thereby keeping each other under control. The result of this was, inter alia, that supranational organizations, such as the United Nations, were merely instruments for managing this East-West conflict. Foreign and security policies were therefore dictated to a great extent by military considerations. Little, if any, need was seen to more extensively shape political processes.

This no longer holds true after the Soviet Union and the Warsaw Treaty Organization disintegrated, allowing the world to return to a 'complex normality'. Acute conflicts were no longer restricted to North-South discord or to antagonisms between developing and industrial countries, but included, in addition, a generation (present-future) conflict.

Meanwhile, the world appears to be at risk of sinking into social, economic, ecological, political, and military chaos. Today, supranational organizations either possess a limited ability to bring the world to order or have ceased to exist altogether. Nation-states in competition with each other increasingly attempt to carry through their national interests to the disadvantage of their neighbours. The principle of 'the right of the stronger one', and thereby the use of war as a viable means to force through regional interests, more and more seems to be becoming an attractive foreign policy option.

One area of regional conflict that is of global importance is the access to fresh water. While merely a supply problem in industrial nations, fresh water is an independent production constraint in arid and semiarid regions. In the future, it will not only be a limiting factor in the production of food, but also a key to the development of industry. Hence, fresh water will become a major element that defines national security policy. Thus it is necessary to expand the definition of the term 'aggression' to include the deliberate withholding of water from a state.

The new situation, however, in contrast to the Cold War period, also opens up new areas and possibilities of development. It constitutes a great challenge to the politics of gaining initiative in complex fields as well as the process of actively shaping processes.

Nation-states will continue to remain the determining factor, though, due to the increasing number of transboundary problems, they are sometimes no longer in a position to exert their sovereignty.[2] They will, none-the-less, decide which international legal norms are implemented and which are not.

---

[2] Cf. Epiney 1995, p.310.

International politics must adapt to the fact that it will only partly be influenced by supranational organizations. In the future it will to an increasing extent be shaped by bilateral or multilateral networks of interconnected relations that are highly complex. In essence, nation-state decision-making criteria will be suppressed in favour of promoting regional policy. This will require:

- Competent and flexible diplomacy that recognizes and utilizes possible package deals in an inscrutable network of causes and effects involving diverse challenges and risks;
- The definition of a new relationship between politics and science, as this is the only way to develop effective preventive and/or adaptation strategies as well as integrated political approaches;
- The provision and appropriate use of communication (consensus, cooperation), as well as power projection instruments (confrontation, intervention).

Above all, current and future challenges necessitate investments in the establishment of civil societies[3] which are characterized by their internal integration and control. This will produce reliable cooperation partners with regard to their external relations.

Regarding conflicts over water resources, a policy must be developed that incorporates the greatly varying interests, levels of development, social conditions and abilities of users. Such a policy must allow the respective users to assert themselves, while generating cooperative agreements that result in the fair distribution and optimal use of this resource without destroying its ecological foundations. In this respect, international law is *one* instrument that can provide reliable rules for successful transboundary water management.

In the Draft Articles on the Law of the Non-navigational Uses of International Watercourses[4] elaborated by the International Law Commission (ILC) on behalf of the General Assembly of the United Nations and debated in October 1996 and March/April 1997 in New York, the following question is to be analyzed: Can international water law be a promising political instrument to settle conflicts of interests?

It must be pointed out here that the topic discussed in this paper will be placed in a broader context, in that all efforts to solve the water problem will amount to nothing if immediate measures are not taken to counter the various causes of uncontrolled population growth. This is the *significant factor* regarding the long-term assurance of world-wide water supply.

---

[3] For the understanding of the term 'civil society', which means more than 'democratization', see, e.g. Boutros-Ghali Boutros 1996: An Agenda for Democratization, New York; Lietzmann, Hans 1993: 'Verfassungspatriotismus' und 'Civil Society', in: Voigt, Rüdiger (ed.): Abschied vom Staat, Rückkehr zum Staat? Baden-Baden, pp.205-227; Norton, Augustus Richard 1993: Peacekeeping, Civil Society and Conflict Regulation, in: Canadian Defence Quarterly, 1, Special, pp.31-33; Steffani, Winfried 1994: Für eine Zivilgesellschaft!, in: Politische Meinung, März, pp. 23-29.

[4] United Nations: Draft Articles on the Law of the Non-navigational Uses of International Watercourses (1991).
See the Annex in this volume.

1.2 The Contradicting Principles of Sovereignty and Integrity

> Water knows no frontiers; as a common resource it demands international cooperation.
> Committee of Ministers of the Council of Europe: European Water Charter, 1967

Natural watersheds do not follow artificially established national borders. As a result, 40-50% of the world's population uses international transboundary basins. Thus water counts among the "common-pool resources, which are partially excludable and rival" and is characterized by the fact that all riparian states endeavour to nationalize (internalize) its use and to internationalize (externalize) the burdens.[5]

Several factors result in varying means in which a nation-state asserts itself in a water distribution conflict. These include:

- Respective position in relation to the basin (upstream state, downstream state, state bordering transboundary waters);
- Access to scientific and technical developments;
- Level of economic, political and, last but not least, military power.[6]

International water law first developed in Europe and North America in the beginning of the 20th century as a young discipline of international law research. A system of regulations was required to manage heavy strains that were imposed on basins due to increased traffic, waste disposal, hydroelectric production, and agriculture requirements. It is no surprise that the development of international water law was closely connected to and mutually influenced by the development of international environmental and economic law. In addition to the UN World Water Conference held at Mar del Plata in 1977, conferences, action programmes and conventions from the latter disciplines have given new impetus setting guidelines and standards to further develop international water law.[7]

It was readily evident that the use and management of single rivers or lakes did not realistically reflect regional water problems. A broader concept was therefore created that defined inland water basins by geographic watersheds and took into account surface as well as underground water flows. Ultimately this was the only way to properly judge the demands of the riparians with regard to optimal water use, and to grasp the full extent of the effects of pollution with respect to a given water body. This new model clearly demonstrates that water cannot be effectively

---

[5] See Benvenisti 1996, p.388.

[6] In 1986 SIPRI summarized the imminent situation as follows: „The potential for future conflict over scarce fresh waters is growing in various regions of the world ... Special problem areas involve upstream/downstream water competition in arid and other regions of rapidly increasing population." (quoted in Westing 1986, p.9).

[7] These - to name a few examples - included the UN environmental conference „Only One Earth", held in 1972, the Declaration and the Programme of Action on the Establishment of a New International Economic Order from 1974, the UN Charta on Economic Rights and Duties of the States from 1974 and the UN Conference on Environment and Development 'Earth Summit', held in 1992.

managed within the confines of a state border, and hence must be governed by international regulations.

Standard literature today still applies three principles of equal importance regarding the judgment of demands on a basin:[8]

- The first principle - absolute *sovereignty*;
- The second principle - absolute *integrity*;
- The third principle - *sovereignty restricted by consideration* for the interests or integrity of other states.

Common sovereignty or the *legal community* consisting of several states bordering a water body is considered to be a fourth principle. However, it was considered to be too difficult to derive concrete legal consequences from a general community of interests, and therefore it does not hold much importance today.

The respective principle that is supported by a specific state depends mainly on its position with relationship to the water source. The first principle of absolute sovereignty is advantageous to the upstream state because it regards water bodies as an integral part of a state's national territory. Thus upstream states maintain the right to utilize the water source in any way that they deem reasonable. Claims by a dissenting downstream state are rejected as interference in their domestic affairs. According to the principle, the upstream state is not subjected to any duties outside the contract with regard to its utilization of the water body. On the other hand, the downstream state prefers the second principle of absolute integrity allowing it to accuse, and perhaps even censure, the upstream state for illegally taking measures

[8] *The Doctrine of Absolute Sovereignty (HARMON Doctrine)*: „Territorial sovereignty is the sovereignty applied to a specific country or basin, the right to make decisions with regard to this country or basin without having to consult other countries and their citizens. (Hefter 1988, quoted in Berber 1955, pp.13).
The Doctrine of Absolute Integrity: "A river which flows through the territory of several states or nations is their common property ... Neither nation can do any act which will deprive the other of the benefits of those rights and advantages. The inherent right of a nation to protect itself and its territory would justify the one lower down the stream in preventing by force the one further up from turning the river out of its course, or in consuming so much water for purposes of its own as to deprive the former of its benefits ... The gifts of nature are for the benefit of mankind, and no aggregation of men can assert and exercise such right and ownership of them as will deprive others having equal rights, and means ... of such enjoyment ... the common right to enjoy the bountiful provisions of providence must be preserved..." (Farnham 1904, quoted in Berber 1955, pp.13).
The Doctrine of Sovereignty Restricted by Consideration: "We have found that when a waterway crosses two or more territories in succession, each of the States concerned possesses rights of sovereignty and ownership over the section flowing through its territory ... A State has the right to develop unilaterally that section of the waterway which traverses or borders its territory, insofar as such development is liable to cause in the territory of another State only slight injury or minor inconvenience compatible with good neighbourly relations. On the other hand, when the injury liable to be caused is serious and lasting, development works may only be undertaken under a prior agreement..." (Memorandum of the ECE 1952, quoted in Berber 1955, pp.13).

that are disadvantageous to its territory. Clearly, when these two principles of international law come into conflict, they do not permit a reconciliation of interests. In the end the right will lie with the stronger state, and as a consequence present and future challenges are not being met in a constructive manner.[9]

This dilemma is well illustrated by the states bordering the Tigris and Euphrates rivers. Turkish officials claim the right to do as they wish with the water coming from their mountains, arguing that they, for their part, do not make any claims on Arabian oil. On the other hand, Iraq demands 60% of the annual water capacity of the Euphrates, based on over 5000 years of history in which water from the Euphrates was available for all of the other riparian states. A similar situation exists on the Nile between Egypt and Ethiopia, where we will shortly have to reckon with a comparable clash of these positions.

1.2.1 The Helsinki Rules

To overcome tensions the International Law Association drafted recommendations for dealing with international water systems during its conference in Helsinki in 1966 and at other follow-up conferences. As a fundamental principle for state practices, it demands that all repercussions resulting from interventions in a 'hydrographic unity' regarding the utilization of basins be recorded, assessed and evaluated. The Helsinki Rules is an essential document that sets the course for international water law. It was founded on the third principle of *integrity* and *restricted sovereignty* and provides further basis for the 'equitable apportionment and utilization-doctrine'. The most important standards set by the Helsinki Rules are:

- The obligation to carry out reasonable compensation measures if the interests of states have been violated;
- The consideration of all relevant technical, economic and financial factors while planning and carrying out a project;
- The requirement to inform all riparians of a pending project;
- The observation of an appropriate waiting period before the construction of a project is started;
- The collection and open exchange of data;
- The arbitration of objections by means of consultative committees and arbitration commissions;
- The rejection of all absolute utilization restrictions or requirements.

Yet, whoever expects that the Helsinki Rules entail binding regulations for all problems and issues relating to the distribution of water, will be disappointed, for they fail to sufficiently define the utilization or consumption of water, establish binding criteria for equitable utilization, and clearly fix priorities with respect to different kinds of utilization. However, they do call upon the riparians of a particu-

[9] With the arbitration decision on Lac Lanoux the term of 'absolute sovereignty' is rejected, see RIAA (Report of International Arbitral Awards), Vol.12, p.281.

lar international basin to engage in bilateral and multilateral negotiations. Furthermore, they require specific obligatory conduct, e.g. consultations, as well as procedures for the conciliation of interests, e.g. commissions, among the negotiating partners. The negotiating partners are encouraged, in accordance with general legal principle, to recognize the rights of the other party. To accomplish this they must forego their positions of absolute sovereignty and integrity when developing water convention or river area commission treaties.

Hence, upon closer consideration, the vagueness of the details proves to be advantageous as each basin has its own characteristics. A framework of conditions and interests that best dictate the actions of the riparian states in particular cases can vary considerably. During negotiations regarding different kinds of utilization the interests of these states are to be assessed and weighed. Issues regarding the consumption and use of water from a basin are settled in accordance with the characteristics of the water body, the population, etc. Objections as well as necessary modifications of utilization shares are resolved by consultation committees and arbitration commissions.

In Central Europe all valid contracts regarding the utilization of international basins have been completed within the basis of customary international law provided by the Helsinki Rules. The Convention on the Protection of the Rhine and the Lake Constance Treaty[10] are exemplary in their considerations of these rules.

However, it is important to note that forms of society have developed which already base their efforts on the conciliation of interests, thus pursuing relative instead of absolute claims. This is true especially in Western and Central Europe, where common actions directed towards mutual advantage have created societies that are closely interlinked. Thus they meet the criteria required of a *civil society*.

The Convention on the Protection and Use of Transboundary Watercourses and International Lakes (Helsinki Convention) represents an advanced development of the Helsinki Rules, which embodies the third principle of *restricted sovereignty* in the international law system of Western and Central European societies. This does not, however, guarantee that distribution conflicts can be prevented, but this instrument does pave the way for the civilized settlement of water conflicts in Europe.

Due to the fact that water conflicts in Europe are generally an issue of quality instead of quantity, European models cannot simply be applied to arid and semiarid regions where water is more a question of survival.

Nevertheless, negotiations over non-European basins, too, have been oriented towards the standards of the Helsinki Rules, even if sometimes only in part. Within the United States the principles of the Helsinki Rules are applied as a basis between the states. Regarding the Indus, India had at first assumed a position of absolute sovereignty. Later, in 1966, India and Pakistan took into account the Helsinki Rules when negotiating contractual provisions. In doing so, they have since acted pragmatically by deliberately leaving out issues of sovereignty and integrity and putting to the core technical and economic issues. For the basins of

---

[10] For the functions and developments of the Rhine-Commission, see Wieriks 1997. For the introduction of procedures leading to 'equitable utilization', see the Lake Constance Treaty, 30 April 1966, Bundesgesetzblatt 1967 II, p.2314.

the Chad and the Zambesi rivers, similar agreements have been concluded between the riparian states and laid down in a master plan.

The conflict of interests over the River Jordan has in the meantime adopted a fundamental new quality. Due to the division plans of the United Nations, up to 50% of the water resources essential for Israel originate outside its national boundaries. The occupation of the Golan Heights was not only an act of strategic military importance, but also crucial in safeguarding Israel's water supply. Now it is recognized that peace in this region can be brought about only through binding treaties agreed upon by equal partners. It still remains to be seen whether the negotiations between Israel, Syria, Jordan and the Palestinians over the sources of the River Jordan will lead to cooperation or to stronger confrontation.

Yet, a course for cooperation on the basis of the Helsinki Rules has been set by the peace treaty between Israel and Jordan.[11] In this treaty the issue of water plays a key role in the future cooperation between the two parties.

The two principles of *absolute sovereignty* and *absolute integrity* more or less exist unchanged in the basins of the Aral Sea, the Nile and the Tigris-Euphrates. Here the Helsinki Rules are referred to superficially in discussion. In regions of acute conflict such as the Middle East and Central Asia, no fixed rules are accepted or applied by the riparians.

The United Nations entrusted the United Nations Environment Programme (UNEP) and the International Law Commission (ILC) with creating a system of reasonable water management and enforceable water law that can be internationally applied beyond Europe and North America as well. Other specialized agencies of the United Nations, such as the United Nations Educational, Scientific and Cultural Organization (UNESCO), the World Bank (IBRD), or the World Health Organization (WHO) have also addressed this issue. The activities of these and other organizations must not be underestimated, as they provide a forum facilitating the exchange of experience, rules, standards, ideas, and national interests. Hence their contribution is essential to the crystallization of international, political, technical and legal standards of practice.

1.2.2 Draft Articles on the Law of the Non-navigational Uses of International Watercourses[12]

In accordance with the General Assembly (Resolution 2669 from 8 December 1970) the ILC has drafted a treaty for discussion that will incorporate the third principle of *restricted sovereignty*, thus binding conflicting parties to negotiation

[11] Treaty of Peace between the State of Israel and the Hashemite Kingdom of Jordan, (34 I.L.M. 46 (1995)): "The Parties, recognizing the necessity to find a practical, just and agreed solution to their water problems and with the view that the subject of water can form the basis for the advancement of cooperation between them, jointly undertake to ensure that the management of their water resources do not, in any way, harm the water resources of the other Party."

[12] For more detailed information concerning the development of the Draft Articles, see Arcari (1997) and Tanzi (1997).

and cooperation. To a large extent the draft takes into account both recent developments in the discussion of the international economic order as well as the efforts for worldwide environmental protection. In its origin it also reflects the basic problem of international law. The efforts of the ILC in 1980 to define an 'international' water basin as a 'shared natural resource' did not find general acceptance. The same holds true for more concrete definitions of 'equitable utilization' or 'appreciable harm'.[13] The current draft treaty merely represents 'a smallest common denominator' which establishes principles for the dealings between states within an international basin, but is open for further concrete developments. Its objective is to:

- Strengthen sensible conflict management procedures using customary international law as a basis, with the possibility of transforming resulting agreements into contract law;
- Create efficient regulations by contractual binding of the states;
- Require an accurate accounting of the adverse effects of their sovereign actions;
- Enforce international laws and regulations;
- Shift the emphasis from the principle of sovereignty to the principle of solidarity.

The draft was discussed controversially in the ILC sessions of 3 May - 23 July 1993 and in the session of the Sixth (Legal) Committee of the General Assembly on the basis of the ILC report in November 1993. The discussion centred on the issue „whether the final outcome of the commission's work should be a framework convention or model rules".[14] The statements made about the drafts clearly reflect the endeavours of most states to prevent restrictions with regard to the planning and execution of national development programmes. With each binding regulation there lay the risk of intervention in their economic, agricultural and industrial pol-

[13] Benvenisti 1996, pp.399.

[14] 84th General Assembly, 6th Committee, 28th Meeting: Press Release GA/L/2799 of 5 November 1993. United Nations (1993, pp.74): „Most of the members who commented on this article found it useful particularly if the article were to become a framework convention whose provision were meant not to be binding but to act as guidelines for States."
India explains: „Account should be taken of the principles of State sovereignty over uses of natural resources... Equitable utilization of watercourses did not mean equal utilization. Optimal utilization of watercourses depends on States' resources in light of their needs. Significant and substantial harm to watercourses should be avoided, but the developmental needs of States should be respected." Although Pakistan could be affected by this Indian policy, it also supports the desire for non-commitment, but adds: „There must be a balance between upper and lower riparian States."
Turkey explains: „The affected State should not necessarily have the power to veto activities ... The State of origin might not be expected to refrain from activities if those activities were fundamental to development of the country...Everything would depend on the relationship between the State that had committed an act and the affected State." Thailand: „A balance must be struck between the sovereign rights of a State and the community of interests. That task was perhaps best left to specific agreements ... should be given precedence over a framework convention that the Commission might produce."

icy. Thus the states of Argentina, Bahrain and Ethiopia preferred a non-committal framework convention. Turkey and Thailand did not directly express their opinion on the issue of required framework, but from their general statements it could be concluded that they preferred a convention that was even less binding. The only state advocating a 'legally binding agreement' was South Korea. The question as to the application of a possible convention to existing treaties met with broad disapproval.

Hence the ILC was faced with the dilemma that several states, especially those in problem regions, would not accede to a treaty containing binding agreements with respect to the utilization of water. On the other hand, there was general agreement that a Water Convention should entail procedures for conflict management. In this respect the majority demanded a regulation that dictated the use of political and diplomatic channels instead of legal procedures. The discussion repeatedly referred to the Helsinki Rules as being exemplary, in that they allow through negotiation the establishment of regulations without the creation of binding obligations. The Niger, Gambia and Rhine rivers were cited as examples in which regulations were established without the framework of a Water Convention.

In July 1994 the second reading of the draft ('final text') was completed.[15] Since it was to be translated into a framework convention, the draft generated positive reaction from the majority of member states and only required a few supplements. Thereby the demands for the establishment of conflict management procedures were supplemented as Article 33. In the event of the failure of consultations and negotiations, the draft provided for a fact-finding machinery. One fault of this mechanism was that it did not become more strongly institutionalized. The proposal to include in the convention transboundary ground water resources, independent of their connection to a basin, was not realized. The fears that this would lead to a renewed discussion and endanger earlier achievements were too great, leading to a recommendation later added to the draft, the value of which, however, was very restricted.

The General Assembly of the United Nations followed the final recommendation of the ILC to hold an international conference on the basis of this draft. The resolution of 9 December 1995[16] mandated this conference to be convened for the period from 7-25 October 1996.

[15] Cf. United Nations 1994, pp.195.

[16] United Nations General Assembly: Report of the International Law Commission on the work of its forty-sixth session, A/49/738 from 2 December 1994, pp.15: „The Sixth (Legal) Committee shall convene as a Working Group of the Whole open to all States Members of the United Nations or members of specialized agencies to elaborate a framework convention on the basis of the draft articles."

1.2.3 Negotiation of the ILC Draft in the Legal Committee of the General Assembly[17]

It was the intention of the conference leaders[18] to pass, if possible, the existing ILC text as a 'balanced package' giving equal consideration to the interests of the different riparian states. This, however, had already failed at the beginning of the working party sessions. A large number of states demonstrated their disagreement with the existing text and insisted on further discussions regarding fundamental as well as specific issues. Dissatisfaction was already evident in the preliminary stages of the negotiations, during which a great number of written comments and requests were delivered.[19] On 18 October an open outburst occurred which continued throughout the third week of negotiations. The confrontation could not be quelled and affected practically all articles. Furthermore, it exposed a definite trend in which a large number of states were unwilling to make any compromises. Contrary to the intended spirit of the document under discussion, the individual contributions clearly reflected the explicit interests of the particular states concerned. Therefore, the conference had no chance of reaching a successful conclusion.

The core of upstream riparians comprised Turkey and Ethiopia. However, supporting them were India, China, Switzerland, Slovakia and, clearly discernible, France too. Together they endeavoured to change the substance of the existing draft in such a way that it became even less binding. They called for a 'framework convention' and disputed the detailed regulations of the existing draft as being excessive. In addition, on the last day of negotiations, there was an attempt to deliberately delay as long as possible the date that the convention would go into effect, so that the existing upstream utilization projects could proceed without hindrance. The failure of the conference was, especially for Turkey and Ethiopia, tantamount to the political success of the upstream states.

The motives of France, cooperatively integrated in Europe in the Rhine Commission and therefore aware of the value of such conventions, can only be speculated as follows: first, France is a big foreign investor in the Southeastern Anatolian Project in Turkey. Secondly, there is a good chance for French industry to repeat this success in Ethiopia as soon as the plans for the damming and utilization of the River Nile are realized. The favourable geographical position of Djibouti enhances France's chances of getting 'its foot in the door'. It can therefore be concluded that France's economic interests had been decisive.

---

[17] For a short summary, see McCaffrey 1997.

[18] Chairman of the Working Group of the Whole: Prof. Chusei Yamada, Chairman of the Drafting Commitee: Prof. Hans Lammers, Expert Consultant: Robert Rosenstock.

[19] Report of the Secretary-General: Comments and Observations received from States, A/51/275 and Add.1-3, August-October 1966. Turkey criticized the draft as, in its opinion, it exceeded the scope of a 'framework convention': „However, they clearly include provisions which go far beyond the scope of a framework agreement ... it would be necessary to rearrange section III and IV of the draft articles so as to avoid too specific and detailed provisions." (A/51/275, p.16).

The group of downstream states consisted of Syria, Portugal, Egypt, the Netherlands, Iraq, Brazil and Hungary. Their main objective was to raise the environmental standards and press for closer integration in issues of consultation and rules of conduct. These higher value political objectives that were represented by the interests of the downstream states were primarily supported by Finland and in part by the USA, Canada and Germany. The Finnish delegation is to be commended for its competence, expertise, and presentation of extensively prepared ideas that were comprehensible and to the point. In this respect, it is most regrettable that the Finnish delegation received insufficient support, even among the European states.

The conference ended on 25 October 1996 without generating an agreement on a convention. It was precisely in the most fundamental issues that no consensus was reached. The discussion of Articles 7 and 33, which contained key elements, remained open. An important consequence was that no rules for peaceful conflict settlement exist. This was not only problematic for regions such as the Nile Basin and the Euphrates-Tigris Basin, but also of significant concern for Europe. The failure of the conference sent conflicting signals in view of the trend toward stricter observance of environmental objectives and promotion of legal relations in conflict resolution.

The General Assembly was, on the basis of the work that was achieved, given a recommendation to reconvene the conference as soon as possible in order to complete the convention. The conference reconvened earlier than expected on 24 March and concluded on 4 April. However, it appeared once again that a consensus would not be reached. Indeed, many delegations that asserted opposing positions during the 1996 conference, assumed much more moderate roles, in that they were persuaded not to oppose the general text of the convention although they did not always agree with specific parts of it. This, however, did not apply to all such states.

On the last day of the conference, the committee chairman decided, contrary to common practice in the UN today, to vote on those articles that were contentious. Thereafter the committee voted on the convention text in its entirety. The result of this vote was:

For	42 nations	upper and/or lower riparian states
Against	3 nations	China, France, Turkey
Abstentions	18 nations	mostly upper and a few lower riparian states, e.g. Egypt, India, Israel, and Pakistan

Strong opposition came from Turkey and France. In the French view numerous procedural violations were committed in order to gain acceptance of the convention text[20] and therefore the adoption of the text was not in line with the rules. Turkey accused the committee of generating a document whose terms were more stringent than a framework convention, and therefore overstepped the bounds of the initial General Assembly mandate. Furthermore Turkey asserted that it would not accept any customary international law standards that resulted from this text. Turkey's position forecasts that a cooperative solution to the conflict concerning the Tigris and Euphrates water system cannot be expected, and in the future confrontations with affected nations are possible.

The French position is a curious one as far as the European Union is concerned. Other than obvious economic motivations, it can be assumed that the French objective is to minimize the codification of international law so that France can pursue its own agenda on the world's foreign policy stage with maximum flexibility. China is similarly motivated. In the General Assembly vote, France, fearing it would lose support from the EU countries in other matters, abstained. Such behaviour indicates that Europe is far removed from developing a joint foreign and security policy, although European agreements concerning the use of its own international water systems can serve as example in this matter.

In the view of Germany, the following point can be considered a success: originally, the regional river commissions had merely a technical character in the convention in that they were anchored in Art. 24, which emphasized 'management'. Through German initiatives, supported by a large number of co-sponsors, consensus was gained in incorporating them into Art. 8 as well, which emphasizes the 'general obligation to cooperate'. Thus the regional river commissions will in the future take on a new political dimension.

Once more it is clear that it ultimately depends on the will of the states if the creation of an international water convention allowing the peaceful settlement of basin conflicts is to succeed. Many of those concerned are obviously unwilling to ratify a convention in order to preserve their perceived vital national interests.[21] The international community requires strong political and economic partners who will emphasize the importance of UN efforts and who are in a position to contribute their experience and propose package deals. These partners have the ability to create framework conditions in which voluntary solidarity could develop instead of

---

[20] See the annex for extracts of the Convention text and a short summary of the results of the discussion in the committee. The final state of the discussion in October 1996 can be found in the following UN documents: Report of the Drafting Committee: A/C.6/51/NUW/WG/L.1 with Corr. 1 and 2, L.1/Add.1, L.1/Add.2 with Corr.1, L,1/Add.3 and Add.4; Proposals: A/C.6/51/NUW/WG CRP.2, CRP 3, CRP 7 regarding the preamble; A/C.6/51/NUW/WG CRP.42, CRP 68, Amendments from Austria, Canada Portugal and other states to Article 7; A/C.6/51/NUW/WG CRP.71, CRP.62/Rev.1, Proposals from Syria and from the Chairman of the Drafting Committee regarding Article 33; Summary Record of the 12th Meeting: A/C.6/51/SR.12 to 24. The final state of the discussion in March/April 1997 can be found in the following UN documents: Report of the Working Group of the Whole: A/C.6/51/NUW/WG/L.3 with L.3/Add.1, L.4 with L.4/Add.1; Report of the Sixth Committee of the General Assembly: A/51/869, 11 April 1997; United Nations General Assembly: Resolution 51/229, 21 May 1997.

[21] Cf. Heintze 1997, p.297.

the mere implementation of a convention. These nations are able to bear the costs of conflict settlement procedures. This point was repeatedly emphasized throughout the conference by developing countries who often base their rejection of UN concepts solely on the fear of being unable to carry the economic burdens imposed by international agreements.

It is important to note that the water convention of the UN does not stand alone; it is, rather, integrated in a network of already existing and developing regulations and a number of other activities.[22]

1.3 Enforcement of International Law

> The global environmental problems pose three specific challenges: setting the rules, monitoring and verifying compliance, and providing an authoritative and binding method of settling disputes.
> Geoffrey Palmer, former Prime Minister of New Zealand, 1992, p.283

National law can generally be set and enforced when violated. The existence of international laws does not guarantee the sanction of conduct associated in cases of non-conformity with a contract. Decisions of an arbitral tribunal require that the conflict adversaries themselves bring the case before the court. The readiness to do so, however, has receded since the end of World War II.[23] This indicates clearly that the majority of states are not disposed towards compulsory international jurisdiction. Even if judicial proceedings result in arbitral decisions or judgments, there are ways in which these can be avoided. When submission to jurisdiction has been declared, the principle of good faith rules, e.g. an arbitral decision, by no means includes the right to enforce compulsory acceptance. Response to non-compliance of a decision can at most include repressive measures short of military actions. Therefore the benefits of an arbitral tribunal, especially for the weaker party that has no other means of gaining recognition for its interests, are most doubtful.

A UN World Council for Environmental Issues as proposed by Bächler[24] may well provide an interesting model. He proposes that, on issues regarding the environment, the Security Council be authorized to invoke Chapter VII measures (Enforcement) in accordance with the UN Charter. In this connection it must be remembered that measures according to Chapter VII presuppose military aggression according to the UN definition of aggression.[25] The present definition, limited

[22] E.g. the Convention on Climate Change and the Montreal Protocol of 1992; the Convention to Combat Desertification of 1994; the ILC draft regarding the liability of the states, by which the 'polluter pays principle' is to be embodied in the law of contract.

[23] Cf. Ipsen 1993, pp.955; Schweitzer 1985, pp.759; pp.84.

[24] Bächler 1993, pp.146.

[25] 'Force' in the sense of Art.2 of the UN Charter means military force. Measures according to Chapter 7 presuppose military aggression as legitimation. The resolution of the UN General Assembly about the definition of aggression dated 14 December 1974 (printed in: Schweitzer 1985, pp.740) says in Article 1 inter alia: „Aggression

to armed force, is in today's context too restrictive, as many states are able to threaten the existence of others in a manner that does not require the use of military might. The saying 'to leave someone high and dry' can be taken literally with respect to the water problem. If the international community is to be allowed to intervene in cases of non-military aggression, it is necessary to redefine 'aggression' in order to justify intervention.

This gives rise to the question of how the UN can obtain the consensus of the international community in order to amend the Charter and create an Environmental Council.[26] One must take care to ensure that such a council generates a solution to the problem rather than becomes a problem in itself.

Herewith, however, support should not be rendered to those who seek to replace international law by a mere balance of power. International law, arbitration tribunals and courts of justice are worldwide and extremely important in that they can make significant contributions to modern security architecture. In a world community where the division of labour is of utmost importance, the discovery of a breach of contract has a decidedly negative effect on further economical and technical integration. In the end, it is not military retribution that the states fear in the event that they violate or can not fulfill international treaties, but rather, fear of isolation from the international community, which could have an adverse affect on the development of their own economies.

The question today is therefore not how a perfect system of control mechanisms and institutions can be constituted, but rather how customary international law, which is already based on relatively successful procedures, can be converted into binding rules that resolve conflicting interests through negotiation rather than armed encounters.[27]

The answer is, ultimately, that an international policy must be pursued that devaluates the principle of sovereignty. It must become clear to nation-states that international solidarity is advantageous. In this context it is fundamentally important that negotiations, first and foremost, deal with the utilization and not the possession of water. The question of possession is nothing more than a question of power. It upholds sovereignty and maintains an atmosphere that in the end generates conflict. If the issue of possession remains in the centre stage, water may

means the use of armed force by one state against the sovereignty, the territorial integrity or the political independence of another state..."

[26] „States are often uneasy about efforts to codify and progressively develop rules of general international law on a subject since they fear the result could prejudice their position in a current or prospective controversy." (McCaffrey 1993, Vol. 45, p.95).

[27] „International society is only slowly waking up to consequences of its own actions toward the global commons. The challenge is how to legislate temperance using the only means at our disposal, international law (...) To deal effectively with the global problems, a form of legislative capacity is essential. Rules that are binding on nations will have to be made by means other than unanimous consent (...) The argument here is not for some utopian system of world government. It is merely for a limited extension of the existing institutions of international law so that the law can cope effectively with a new problem. The proposal does require nations to surrender some sovereignty." (Palmer 1992, pp.282).

eventually no longer be the root of a conflict, but rather an instrument used to force a settlement in a completely unrelated conflict.

Since many states are driven by profit maximization, it is necessary to connect the surrender of sovereignty rights to clearly tangible gains. This advantage need not necessarily lie in the same field of politics, but rather it can embrace the whole range of package deals. In this way a state's acceptance of the international Water Convention, bilateral or multilateral assurances, cost takeover statements in concert with technical, economical and social cooperation can be adjusted to its interests. Linking most favoured nation treatment status to such actions provides a significant incentive by linking the state to powerful coalition of partners in the international community.

The Convention on the Law of the Non-n-Navigational Uses of International Watercourses in essence meets these requirements. It defines, however, as a 'framework convention' the initial minimum standards. On the other hand, it is general enough to allow equal negotiations between parties. Thus these parties are not only capable of achieving a short-term balance of interests, but they can also continue communication to further decision-making, confidence-building and eventually construct mutual dependencies. Herewith they heed past experiences that demonstrate that negotiations between concerned parties are seldom drawn out and that their results are often more stable than solutions imposed by third party intervention.[28] In this manner a winner-loser scenario is avoided, contributing greatly to the success of an agreement. Furthermore, the retention of sovereignty and integrity is largely unquestioned and the negotiations are free of ideological pressures.

Therefore, a 'framework convention' can be considered as intensely political. The content of the Convention leaves the issue of international water management to a large degree open. Such discussions are left to members of Regional Basin Conventions and River Commissions, which are entrusted to define the complex 'individual' rules that are adapted to the particularities of a basin and its further utilization.

1.4 Water Management, Environmental Impact Assessment and Cost-Benefit Analysis

> Clearly, development must be environmentally sound to ensure that the development process does not destroy the resource base upon which future growth depends. This is a fundamental tenet of development.
> Mostafa Kamal Tolba, Executive Director, United Nations Environment Programme[29]

Effective water management coordinates and records all actual and planned utilization of an entire basin according to their type, extent, disposal controls, dis-

[28] „(...) most water-related disputes have been resolved by negotiation and not by litigation..." (Benvenisti 1996, p.400).

[29] Quoted in Thanh 1990, p.ix.

charge and water recycling efforts. Additionally, it ensures the conservation of ecological foundations required for regeneration. It constantly compares supply and demand in order to establish and implement priorities for the users. In doing so, it utilizes all technical resources available.

The first priority of water management is not an increase of water supply, but rather the economic and efficient utilization of the water resources available. The vast array of modern technological tools available to address the problem are but *one* important aspect of water management. Similar importance must be attached to the political, economic, ecological, social and legal aspects of water management. The prospective gain through interventions in existing ecosystems are offset to a certain degree by the pollution created by the intervention itself. If the resulting pollution is irreversible, new environmental balances will result, thereby changing or even recreating the effected systems. Such intervention can quite possibly create environmental conditions (e.g. desertification, salinization of arable land, greenhouse effect, dying forests etc.,) that are even more detrimental to mankind. Consequently, environmental constraints will limit water utilization to a level that is distinctly below that which is technically possible.

It is, today, widely accepted that hydraulic engineering projects are preceded by an examination of their environmental impact. Such studies determine the stability of the ecosystem and thereby the maximum allowable level of pollution that can be created by the project. In addition to this, a cost-benefit analysis, independent of the environmental compatibility examination, is conducted. This will often reveal many a major subregional project as economically impractical. A prerequisite for this analysis, however, is the calculation of 'shadow prices', a step that is often avoided when a prestige object is involved.

The World Bank has learned from this and in the meantime has made available to future borrowers a catalogue of conditions that includes offers of cooperation with respect to hydraulic engineering.[30] It is oriented towards the decision-making criteria which were laid down in the „Plan of Action of Mar del Plata" during the UN World Water Conference in 1977. This criteria, which has been continuously updated, includes the following:

- From the onset each planned project considers possible alternatives. In doing so, socio-economic consequences, potentials for conflict both inside and outside the region concerned, and efficiency are considered.
- Each project is a regional and international cooperative venture.
- Each project plan is preceded by a regional plan which provides information on resettlement, jobs, water supply, disposal of sewage, rights and demands of the users, monitoring systems, health, practical training, and environmental education.
- The people concerned should be allowed to participate publicly in the planning and information gathering process.[31]

[30] Cf. Frederiksen 1994.

[31] Concerning the role of non-governmental organizations, see Brunnée 1997.

It is clear that water management is intricately bound to the scientific process. Such a process of careful collection, management and updating of relevant data makes it possible to make an accurate prognoses for the long-term utilization of a basin. This in itself will generate an objective atmosphere and minimize ideological tensions, thus significantly contributing to negotiations.

Today many states, due to serious domestic difficulties or considerations of prestige, still ignore the criteria set by the „Plan of Action of Mar del Plata".[32] In addition, governmental and non-governmental organizations are gaining increasing importance in the planning of hydraulic engineering projects. They are assuming more and more the role of a control authority mediating between interest groups and competent points of contact in the decision-making process[33].

1.5 Conclusion

Distribution conflicts concerning resources may, but need not necessarily, lead to military settlement. The „Convention on the Law of the Non-navigational Uses of International Watercourses", despite several shortcomings, would, as a 'framework convention', be a valuable contribution to a modern security architecture. The essential actions of the international community would then be to further develop this framework. Contrary opinions assert that since the convention can be characterized as 'soft law', i.e. it is not enforceable, and therefore it is not helpful. On the other hand, what matters is to actively occupy political and diplomatic fields and to pursue political objectives developed in that process.

In developed and developing countries, it must, above all, be made sufficiently clear that in issues of 'ecological security', international political debate and constructive cooperation are required. In this respect it is neither a matter of calling for asceticism nor recrimination, but rather a call for political participation and implementation of principles already recognized as being useful. A first step to success is the recognition of this global challenge, for:

> Technologically, man is prepared for meeting the water shortage. Unfortunately, it is uncertain whether the governments in the world and the authorities are bold enough to grasp the immediate urgency of the problem. If man fails to initiate the measures today

---

[32] "UNEP was established to act as a focal point for environment action and coordination within the United Nations. It lacks any formal powers. Under its Executive Director, Dr. Mostafa Tolba, it points out environmental problems to nations and suggests solutions. In fact, it does much more than its limited powers suggest. It has become an agency that sets out to produce concrete results in terms of treaties negotiated. UNEP can push states, probe their policies and plead with them; it cannot coerce them. UNEP lacks teeth. It has no executive authority. Partly for this reason, UNEP has made generous use of 'soft law' instruments in the international consensus building that it engages in." (Palmer 1992, p.261).

[33] Frederiksen 1994, pp.225.

> that tomorrow provide him with sufficient freshwater, this failure will in the end be a mistake of the leadership, and of the leadership alone.[34]

Therefore a 'global change programme' that anticipates crises and reacts accordingly, rather than one that manages a crisis that is already at hand, is called for. This requires the initiative of industrial nations. Only they possess the relevant technical know-how and the financial power to do so.

In the future, the basic precondition for success is the increased use of package deals to resolve conflicts. This requires that regional policy strategies be connected to water issues in order to deliberately trigger cooperation. This further requires diplomacy supported by extensive economic, ecological, historical and cultural knowledge, as well as further development and crystallization of 'civil societies'. Only under these conditions will the stronger party of a conflict perceive the benefits of complying with the contracts that restrict his freedom of action. In other words, the stronger party will no longer be permitted to enforce its position without constraints. Hence it will come to realize the long-term advantages of compliance and will perceive that the restriction of sovereignty in the end serves its own interest.

Science as well as governmental and non-governmental organizations can contribute significantly by shedding objective and human light on conflicts. The connection of scientific dialog with politics must be strongly supported. Member states of the European Union, as riparians of many different basins, having already benefited from several water management treaties, have extensive expert knowledge which can be provided to other states in problem regions. Furthermore, Europe, as strong economic partner, is in a position to make financial contributions which weaker economies in the region cannot provide. Such contributions should be considered as future investments.

Such conditions combined, if necessary, with assurances to the riparians, can make it quite attractive to relinquish absolute sovereignty in favour of bilateral or multilateral water conventions. This, however, can only be realized if Europe no longer judges security to be primarily a question of relative military power. It must be accepted that security issues include an ecological dimension as well, so that investments set up to further develop 'civil societies' can be politically substantiated.

[34] Smith 1978, p.224.

References

Arcari, Maurizio 1997: The Draft Articles on the Law of International Watercourses Adopted by the International Law Commission: An overview and some remarks on selected issues, in: Natural Resources Forum, August (in print).

Bächler, Günther et al. 1995: Umweltzerstörung - Krieg oder Kooperation? Ökologische Konflikte im internationalen System und Möglichkeiten der friedlichen Bearbeitung, Münster.

Barandat, Jörg 1995: Wasser. Regionaler Konflikstoff weltweiter Bedeutung, Hamburg.

Benvenisti, Eyal 1996: Collective Action in the Utilization of Shared Freshwater: The Challenges of International Water Resources Law, in: American Journal of International Law, 3, pp.384.

Berber F.J. 1955: Die Rechtsquellen des internationalen Wassernutzungsrechts, München.

Brunnée, Jutta; Stephan J. Toope 1997: Environmental Security and Freshwater Resources: Ecosystem Regime Building, in: American Journal of International Law, 1, pp.26-59.

Epiney, Astrid 1995: Das "Verbot erheblicher grenzüberschreitender Umweltbeeinträchtigungen": Relikt oder konkretisierungsfähige Grundnorm, in: Archiv des Völkerrechts, 3, pp.310.

Frederiksen, Harald D. et al. 1994: Principles and Practices for Dealing with Water Resources Issues, World Bank Technical Paper Number 233, Washington D.C.

Heintze, Hans-Joachim 1997: Wasser und Völkerrecht, in: Barandat, Jörg (ed.) 1997: Wasser - Kooperation oder Konfrontation, Baden-Baden.

Ipsen, Knut 1993: Völkerrecht, München.

McCaffrey, Stephen C. 1993: The Evolution of the Law of International Watercourses, in: Austrian Journal of Public and International Law, Vienna, pp.87.

McCaffrey, Stephen C.; John R. Crook 1997: The United Nations Starts Work on a Watercourses Convention, in: American Journal of International Law 2, pp.374.

Palmer, Geoffrey 1992: New Ways to Make International Environmental Law, in: American Journal of International Law 2, pp.259.

Schweitzer, Michael; Walter, Rudolf 1985: Friedensvölkerrecht, Baden-Baden.

Smith, Norman 1978: Menschen und Wasser, München.

Tanzi, Attila 1997: Codifying the Minimum Standards of the Law of International Watercourses: Remarks on Part one and a half, in: Natural Resources Forum, May (in print).

Thanh, N.C.; Biswas, Asit K. (eds.) 1990: Environmentally-Sound Water Management. Bombay, Calcutta, Madras.

United Nations 1991: Draft Articles on the Law of the Non-navigational Uses of International Watercourses, in: International Legal Materials, 30 I.L.M. 1575.

United Nations 1993: Report of the International Law Commission on the Work of its Forty-Fifth Session, New York (General Assembly Official Records, Forty-eight Session Supplement No.10).

United Nations 1994: Report of the International Law Commission on the Work of its Forty-Sixth Session, New York (General Assembly Official Records, Forty-Ninth Session Supplement No.10).

Westing, Arthur H. 1986: Global Resources and International Conflict. Environmental Factors in Strategic Policy and Action, Oxford, New York.

Wierks, Koos; Schulte-Wülwer-Leidig, Anne 1997: Integrated Water Management for the Rhine River Basin, From Pollution Prevention to Ecosystem Improvement, in: Natural Resources Forum, May (in print).

2 International Water Agreements: A Comparative View

Manuel Schiffler

Treaties on the utilization of international watercourses[1] are a key element in the resolution of international water conflicts. But it is precisely in the Middle East, where water is scarce and many countries are dependent on international rivers, that, until now, only a small number of international water agreements have been signed. The few existing agreements display, in addition, an important weakness: they have, without exception, been concluded on a bilateral basis, without including all of the states located along a given river, as is recommended by customary international law.

The process of negotiating further bilateral water agreements and the possibility of transforming bilateral agreements into multilateral agreements at the level of the catchment basin will no doubt be a protracted and difficult one. This process could certainly benefit from experience with existing agreements both in and outside the region. The present paper is an attempt to systematically evaluate these experiences.

2.1 Water Agreements Analyzed

At present there are five international water agreements in force in the Middle East. Some of these treaties are devoted exclusively to water issues, while others deal with water issues and other questions. Arranged in the order in which they were signed, these treaties are:

- The Agreement on the Full Utilization of the Nile Waters, 8 November 1959;
- The Protocol of Economic Cooperation between Syria and Turkey, July 1987;
- The Jordanian-Syrian Agreement on the Utilization of the Waters of the Yarmuk River, 3 September 1987;

[1] According to the terms of the UN Water Convention (Articles 1 and 2), a transboundary river is an 'international watercourse'. This designation is rejected by the proponents of the Doctrine of Absolute Territorial Sovereignty, who continue to speak only of transboundary rivers. The term 'international watercourses' is defined in the UN Water Convention in such a way as to include both surface water and ground water related to surface water.

- The Syrian-Iraqi Agreement on the Utilization of the Euphrates Waters, April 1990;
- The Peace Treaty between Israel and Jordan, 26 October 1994.[2]

Aside from these five agreements from the Middle East, two additional international water agreements are analyzed for purposes of comparison:

- The Pakistani-Indian Agreement on the Indus Water, 19 September 1960;
- The Chloride Agreement [*Chlorid-Abkommen*] between the riparians of the Rhine River, 1976.

The framework of this paper provides no leeway for any detailed consideration of the political, historical, hydrological, and economic aspects of water management in all of the river basins covered by it. For these aspects, the reader is referred to the comprehensive literature on the topic.[3]

2.2 Analytical Matrix

A six-aspect analytical matrix is used to assess the experiences made with international water agreements:

- Consideration of customary international law, in particular the 1997 United Nations Convention on the Law of Non-navigational Uses of International Watercourses (referred to below as the Water Convention) and the 1966 Helsinki Rules of the International Law Association (ILA);
- Consideration of the water rights of other states located on a river, which are not signatories of international water agreements;
- Distribution of the risk of low flow rates;
- Water quality and protection of wetlands;
- Elaboration of procedures and creation of commissions to monitor and enforce treaty provisions, and
- An implicit or explicit linkage of the water question with other (political) issues.

---

2 The 1995 Oslo B Agreement between Israel and the Palestinian Autonomy Authority is not considered here, as it is only an interim agreement. The Nile Agreement, the Jordanian-Syrian Agreement, and the Israeli-Jordanian Peace Treaty are available to the author in complete versions; the Turkish-Syrian Protocol is available to the author only in relevant excerpts. The author has made use of secondary sources to assess the Syrian-Iraqi Agreement.

3 For the Nile, Jordan and Euphrates rivers, see the contributions by Schiffler, Dombrowsky, and Scheumann in the present volume, and the literature cited there; for Pakistan and India, see Hofmann 1997 and Lowi 1993, pp.61-67; for the Rhein, see Durth 1996.

2.3 Taking Customary International Law into Account

The most important provision of international law covering international watercourses (apart from navigation) is the 1997 UN International Water Convention. It was adopted by a great majority of the UN General Assembly and is set to come into effect as soon as it has been ratified by 35 states. Also of significance is the 1966 Helsinki Rules of the International Law Association; they correspond in important points to the International Water Convention.[4] Both documents have, it is true, not yet become binding international law. Moreover, the UN Water Convention will become binding law only for the states that ratify it. Ratification by all of the states in the Middle East located in a river basin is unlikely, since at the UN important states from the Middle East voted against the convention (Turkey) or abstained (Israel, Egypt). Still, as far as resolving international water conflicts is concerned, the UN Water Convention and the Helsinki Rules established an orientation framework that can hardly be ignored.

Both documents underscore the 'equitable and reasonable utilization' of international water resources.[5] The Water Convention also underscores the „obligation to cause no significant harm".[6] Both documents implicitly reject extreme positions in international law on water rights: the *Doctrine of Absolute Territorial Sovereignty*, which gives the upstream riparian the right to all water resources originating in its territory, and the *Doctrine of Absolute Territorial Integrity*, which gives the downstream riparian the right to unaltered flows. What is propounded in international law instead is the *Doctrine of Limited Territorial Sovereignty*.

In comparing international law with international water agreements, the following points appear significant:

- The Water Convention specifies that existing agreements not be prejudiced by ratification of the Convention. The signatories are, however, encouraged to harmonize existing agreements with the principles of the Water Convention.[7]
- All states located on international watercourses have, according to the terms of the Water Convention, the right to participate in any such agreement.[8] Bilateral international water agreements containing no reference to the rights of other riparian states appear, in the view of the Water Convention, to be in need of correction (see below).

---

[4] See the paper by Barandt and Kaplan in this volume; see also Biswas 1994; Chenevert 1992; Dellapenna 1995; Inan 1994; Kaswaneh 1995 and McCaffrey 1991, and McCaffrey 1993 on the contents, interpretation, and application of the Helsinki Rules and the Draft International Water Convention worked out in the UN International Law Commission.

[5] Articles 5 and 6 of the Water Convention and Article 4 of the Helsinki Rules.

[6] Article 7 of the Water Convention. On the tension between these two principles, see Dellapenna 1995, p.65; Khasawneh 1995, p.24; McCaffrey 1993, pp.98.

[7] Article 3, Paragraphs 1 and 2.

[8] Article 4 of the Water Convention.

- The precise meaning of the term 'significant harm' is in dispute. In the opinion of a Turkish expert in international law, the 1987 Protocol between Turkey and Syria, for instance, which permits a reduction of flow rates by nearly one half, does not constitute any 'significant harm', and, accordingly, Turkey's actions are in accordance with customary international law.[9]
- Some international water agreements are based implicitly on the Doctrine of Absolute Territorial Sovereignty, which does not correspond to the current practice of customary international law. One treaty that is problematic in this respect is the 1987 Jordanian-Syrian Agreement, which cedes to Syria the exclusive water rights to all tributaries of the Yarmuk River that originate in its territory. Moreover, the agreement fails to consider the water rights of Israel and the Palestinians.
- None of the international water agreements looked into specifies the criteria on which the water resources concerned have been allocated. This makes it impossible to compare them with the criteria set out by the Helsinki Rules and the Water Convention.[10]

2.4 Water Rights of Other Riparian States

According to customary international law and Agenda 21,[11] water should be managed at the level of the catchment basin. It is true that an international water agreement need not necessarily include all states located in a catchment basin, but the water rights of non-signatories in a river's catchment basin should be considered in agreements between other states.[12]

This arrangement is not uncontroversial. Awn Khasawneh, Jordanian member of the UN Commission on International Law, has vehemently criticized the right, anchored in the draft to the Water Convention, of a state located in a catchment basin to be included in bilateral international water agreements.

> [This right] flies in the face of the freedom to choose treaty partners, and it is particularly unsuited to the Middle East (...). In a fragmented area like the Middle East, where political symbolism and nationalism are often more important than substance, entitlement to 'butt' into a water agreement between other states is not likely to receive overwhelming support.[13]

The agreements analyzed deal with the issue of including third-party states in three different ways:

---

9 Inan uses the term 'substantial harm' instead of 'significant harm' (1994, p.231).

10 The UN Water Convention (Article 6) specifies seven criteria, the Helsinki Rules (Article 5) name eleven criteria for the utilization of water resources.

11 BMZ 1995, p.147, Paragraph 18.9: "Integrated water resources management (...) should be carried out at the level of the catchment basin or sub-basin."

12 Articles 3 and 4 of the Water Convention.

13 Khasawneh 1995, p.25.

- First, utilization is defined proportionally, so that additional shares granted to third-party states automatically lead - in terms of absolute quantity - to a reduction of previously defined water rights.
- Second, water rights are allocated on a parity basis by the initial signatory states, so that the initial signatory state with fewer water rights is worse off than it was beforehand.
- Third, no rule is specified for consideration of the water rights of third-party states.

The first group includes the agreement between Iraq and Syria. As the first such treaty in the Middle East, it provides for utilization in terms not of absolute quantities but of shares (58% for Iraq and 42% for Syria). This automatically determines the sharing of any reduction, as a result of higher water use by Turkey, of water use to be borne by Syria and Iraq.

The second group includes the 1959 Agreement on the Nile Waters between Egypt and Sudan. It specifies that negotiations with other Nile states are to be conducted jointly by the two signatory states. The reduction of water rights resulting from such negotiations is to „be deducted from the shares of the two Republics in equal parts“.[14] This rule favours Egypt, which is currently entitled to utilize three quarters of the Nile waters. That is to say, Egypt's share would be reduced, in relative terms, three times less than Sudan's share.

The third group includes all other agreements looked into, with the exception of the Rhine Chloride Agreement, which includes all of the Rhine states. The 1987 Turkish-Syrian Protocol is expressly valid only „until the final allocation of the waters of the Euphrates among the three riparian countries“.[15] The agreement is thus conceived as a provisional arrangement. It does, however, leave open the question as to the share of Iraq's water rights as a third-party state that is to be borne by each signatory. Both the 1987 Syrian-Jordanian treaty on the Yarmuk River and the 1994 agreement between Jordan and Israel on the Jordan and Yarmuk rivers fail to mention the water rights of other states concerned. Both agreements settle the water rights of two states only, although the Yarmuk catchment basin, including the lower Jordan, encompasses four states and the Jordan catchment basin numbers (with Lebanon) five states. This non-consideration of other riparian states potentially implies instability of these agreements.

The 1960 Indus Agreement between India and Pakistan does not include the one third-party state concerned, i.e. the People's Republic of China, on whose territory the sources of the Indus River and one of its tributaries (the Sutley) arise. Since major tributaries of the Indus are located in a remote region of Tibet, in which neither large-scale irrigation nor power generation make economic sense, this circumstance has until now not given much cause for concern.

[14] United Arab Republic and Sudan 1963, Article 5(2).

[15] Beschorner 1992, p.40.

2.5 The Risk of Low Flow Rates

The flow rates in a river fluctuate between seasons and between dry and wet years. The average flows may also fluctuate over the long-run due to climatic changes. Depending on the shape of international water agreements, the risk posed by these fluctuations, in particular that of low flow rates, tends to be apportioned unevenly to the signatory states.

The risk of low flow rates can be taken into account in agreements in the following ways:

- One side may bear the risk alone;
- The risk can be split up by allocating, either fully or minus a specified quantity, individual rivers in a catchment basin to one side or the other;
- The risk can be distributed by agreeing on proportional shares.

The only water agreement in the Middle East that imposes the risk of fluctuating flow rates on one signatory state is the (provisional) Syrian-Turkish Protocol of 1987. It guarantees a fixed, average monthly quantity of water (500 m^3/sec.) to be released at the Turkish-Syrian border. The agreement provides for the possibility that this quantity may not be reached in a given month; in the following month, however, the shortfall must be made up for. The risk of year-to-year fluctuations is borne solely by Turkey. Turkey has far greater water-storage capacities (the Keban, Karakaya, and Atatürk reservoirs have together a storage capacity of 87 billion cubic metres (BCM) as compared with 12 BCM for the Assad reservoirs in Syria), and in Turkey evaporation is three times lower than it is in Syria.[16] Turkey is therefore better equipped than Syria or Iraq to compensate for fluctuations in flow rates. For Turkey, the risk involved is substantial: between 1956 and 1962 flow rates were, for six years, distinctly below average.[17] During the drought of 1991 the flow rate was as low as 199 m^3/sec. Turkey nevertheless discharged 500 m^3/sec., which meant a year's delay in filling the Atatürk reservoir. Turkish sources indicate that for Turkey this entailed losses due to projected power not generated amounting to some US$ 500 million.[18] If these water rights are renegotiated among all three states concerned, Turkey is likely to insist on a more equitable allocation of the risk of fluctuating flow rates.[19]

The second group of agreements includes the Israeli-Jordanian Peace Treaty and the Indus Agreement. The peace agreement between Israel and Jordan guarantees Israel a fixed share of the waters of the Yarmuk River, while the (annually fluctuating) remainder goes to Jordan. The arrangement for the Jordan River itself is just the opposite: Jordan receives a fixed share, while the remaining water goes to Israel. Fluctuations in flow rates are borne by both sides. The 1960 Indus agreement cedes all rights to the three western tributaries of the Indus to Pakistan,

[16] Kliot 1994, p.121.
[17] Calculations based on water-flow data from 1938-80, in: Scheumann 1993, p.252.
[18] Mutlu 1966, p.79.
[19] Scheumann 1993, p.252.

while India is accorded the rights to the three eastern tributaries. In this case the water rights are shared by dividing the catchment basin. This arrangement is easy to monitor and poses no metering problems. But arrangements of this sort are only possible when there are several rivers in the catchment basin that are shared by two states, and are regarded by the parties concerned as each carrying sufficient quantities of water.

The third group includes the 1990 agreement between Syria and Iraq, which provides for a fixed share of the waters of the Euphrates for both signatories, splitting up the risk of fluctuating flow rates between the two signatories in this way. However, both the period of time (year, month, week) stipulated in the agreement for fulfillment of the shares and the basis agreed on for assessing the Euphrates' flows from Turkey are unknown. If the time span is a longer one (e.g. one year), Syria, as the upstream riparian, will be in a better position to make more intensive use of the water in times of high demand, later compensating for reduced flows. A shorter time span would restrict this scope enjoyed by Syria.

And finally, some agreements have settled the problem of fluctuating water flows in a somewhat ambiguous manner. One example is the 1959 Nile Agreement, which contains no clear-cut terms for the case that the overall annual quantity of water is lower than what is provided for in the agreement. Should the water quantity be higher, the agreement specifies that the additional amount is to be utilized by both sides equally.[20] In the case of shortfalls the Permanent Joint Technical Commission provided for in the agreement is required to work out a proposal for an equitable allocation.[21] This has apparently not yet happened, but this could in fact mean that in case of a drought, Egypt, as the downstream riparian, would receive less water. The agreement thus contains considerable conflict potential. In fact the flow rates of the Nile have evidently fallen short of the quantity of 84 BCM provided for in the agreement, an amount arrived at from the average flow rates between 1900-1959. In dry years, e.g. 1984, the Nile conducts only 42 BCM. Egypt has thus far been able to use the Nasser reservoir to compensate for fluctuations in individual years. However, a long-term decrease in the average flow rate, with demand on the rise, could lead to tensions with upstream riparians.

2.6 Water Quality and Protection of Ecosystems

The UN Water Convention and the Helsinki Rules recommend that signatory states avoid pollution that would cause 'significant harm' in the territory of another riparian state (while the Water Convention speaks of 'significant harm', the Helsinki Rules refer to 'substantial damage').[22] The Water Convention furthermore

---

[20] United Arab Republic and Sudan 1963, Article 2 (4).

[21] United Arab Republic and Sudan 1963, Article 4 (1e).

[22] Article 21 of the Water Convention and Article 10 of the Helsinki Rules; see ILA 1966; McCaffrey 1993, p.99.

requires the signatories to take consideration of ecological factors as well as protection of the estuaries and the sea.[23]

In fact, however, only one of the international water agreements in the Middle East considers issues of water quality, and not a single includes consideration of aquatic ecosystems. Increasing urbanization, industrialization, and intensification of agriculture, however, have been accompanied, in the Middle East and in other parts of the world, by growing pollution of lakes and rivers with inadequately treated waste water, salts, fertilizer residues and pesticides. Moreover, the international significance of wetlands, which provide a habitat for fish, migratory birds, and rare plant species, is receiving more and more acknowledgment. Some of the wetlands in the Middle East, including Lake Manzalah and Lake Burullus in the Nile Delta, have been placed by governments under the protection of the 1971 Ramsar Convention. The Middle East also contains other wetlands of international importance that are still without any international protection: for instance, the huge Sudd wetland in southern Sudan, the Huleh area on the Jordan River north of the Sea of Galilee, and the marshlands on the Shatt al Arab in the south of Iraq.[24]

The stipulations of the Israeli-Jordanian Peace Treaty concerning water quality include a provision requiring the signatories to „protect the shared waters (...) against any pollution, contamination, (and) harm“ and to jointly monitor water quality. The agreement also stipulates that municipal sewage and industrial waste water are to be „treated to standards allowing their unrestricted agricultural use“.[25] This passage could be seen in connection with the WHO guidelines on the unrestricted re-utilization of waste water in agriculture, which stipulate microbiological water-quality levels.[26] The WHO guidelines do not mention any other pollutants. The peace treaty also specifies that the quality of the water delivered to Jordan by Israel (from the Jordan River or the Sea of Galilee) must meet the standards that Israel has itself set for water from these sources, i.e. the Israeli drinking-water standard. Compared to other international water agreements in the region, the Israeli-Jordanian Peace Treaty thus contains far-reaching arrangements concerning the quality of water. The Syrian-Jordanian Agreement of 1987 only notes that the signatories „each within its territory, shall take all necessary measures for the prevention or minimization of silt accumulation in the Interstate Reservoir“.[27] The reservoir in question is the projected but never realized Al-Wahda reservoir on the common border. Other pollutants are not mentioned by the agreement.

The other international water agreements in the Middle East, those at least whose texts is known, fail to address questions of water quality and the protection of wetlands. The water needs of wetlands should be settled in international agreements covering not only average annual quantities, but also water quality, and seasonal fluctuations in the flow of water.[28]

---

[23] UN 1997, Article 6, Paragraph 1a and Article 23.
[24] See the data in World Resources Institute 1996.
[25] Peace Treaty 1994, Annex II, Article III.
[26] WHO 1989, pp.36-42.
[27] Agreement 1987, Article 11.
[28] See Lanz 1995, pp.66-77.

The experiences of the industrialized countries in including aspects of water quality in international water agreements could prove useful for the Middle East. The 1976 Chloride Agreement between the riparians of the River Rhine could, for instance, be of interest here. The main problem that was to be settled by the agreement was the discharge of chlorides into the Rhine by Alsatian potash mining operations. These chlorides impair the quality of drinking water and the water used by gardening nurseries in the Netherlands for irrigation.[29] The Chloride Agreement provided for separation and storage of the chloride. The costs were to be shared by the Rhine states. However, shortly after the agreement had come into effect in 1987, the government of the Netherlands refused to participate in funding it. The reason was that the big Dutch gardening companies had gone to court, sued for, and won, a damage compensation. This court decision, which is regarded as a model case, confirmed the polluter-pays principle. At the same time, the Rhine states decided to refrain from codifying their cooperation and their commitments in the form of international agreements. They instead decided to cooperate in the form of far more successful and flexible working programs, e.g. the 1987 Rhine Action Programme.[30]The condition required for this type of cooperation is, however, a great measure of stability in the relations among the states concerned, which is not the case in the Middle East.

Another example likewise shows how the polluter-pays principle can be applied to issues of water quality. In 1944 the USA and Mexico signed an agreement guaranteeing Mexico a fixed amount of water from the Colorado River. However, the quality of the water continuously deteriorated due to return-flows of drainage water from U.S. agriculture. This even threatened the existence of Mexican farming on the lower Colorado River. In 1973 the USA declared its willingness to guarantee Mexico a certain level of water quality and to bear the costs involved.[31] In 1992 the USA completed work on the world's biggest desalination plant for brackish water; its capacity is roughly 100 MCM/a. When mixed with the waters of the Colorado the desalinated water has a salt content of roughly 700 mg/litre, a level acceptable for agriculture. Whether the relatively high costs of US\$ 0.43 per m^3 are economically justified by the benefits to Mexican agriculture, or whether the Mexican farmers might have preferred indemnification in the same amount, is an open question. However, downstream riparians in other catchment basins, e.g. Iraq and Syria, could in the future cite this and other precedents and demand that the polluter-pays principle be enforced.

---

[29] Durth 1996, pp.42-23.

[30] Durth 1996, p.38.

[31] In Minute No.242, International Water and Boundary Commission, in: Murakami 1995, p.81.

2.7 Monitoring and Enforcement of Water Agreements

The monitoring and enforcement of international water agreements is a crucial issue. Only if credible procedures are agreed on for this purpose and institutions are created will water agreements constitute more than a piece of paper.

The UN Water Convention provides for detailed information requirements on projected dams or discharges along international watercourses.[32] Downstream riparians are required to be informed prior to the start of construction measures on a river. They must be granted a period of six months to appraise the plans, and construction is not permitted to start during this period. If negotiations are initiated, the period is extended by an additional six months. An elaborate arbitration procedure is provided for in the case that negotiations are not initiated or get nowhere.[33] A *Fact-Finding Commission* is to be set up for this purpose; it will contain one representative from each of the states concerned and a chairman from a country not immediately affected. If the states concerned cannot agree on a chairman, the UN Secretary General has the power to name one. If this chairman is not accepted, he can form a commission with only one member. It is furthermore possible to call for mediation by a joint water institution or arbitration by the International Court of Justice. The question of how the recommendations of a commission's report or a judgment by the International Court of Justice are to be enforced is, however, an open one. Any state opposing the enforcement of a report's recommendations or a court's judgement would, however, in this way be in conflict with the recognized rules of international law, which could entail negative consequences for its international relations in other areas.

As previously mentioned, the rules of the Water Convention apply, however, only to the states that have ratified the convention.

Nearly all of the international water agreements examined provide for the creation of joint water institutions (should these not already exist)[34] that would take on a great variety of different tasks, ranging from the exchange of data to mediation in disputes. The 1959 Nile Agreement provides for the establishment of a *Permanent Joint Technical Commission* that is to be made up equally of Egyptians and Sudanese and the chairmanship, which is to rotate periodically. In spite of all political tensions between the two states in the past decades, the Commission, headquartered in Khartoum, has continued to monitor the agreement. Turkey, Syria, and Iraq set up a *Joint Technical Committee* in 1982, i.e. at a time at which there was no water agreement between the riparians.[35] The Syrian-Jordanian Agreement of

[32] UN 1997, Articles 11-19.

[33] UN 1997, Article 33.

[34] In the UN Water Convention, Article 8 („Establishment of joint mechanisms or commissions") and Article 24 („Establishment of a joint management system") provide for creation of commissions, but without more precisely defining their tasks.

[35] The 1923 Treaty of Lausanne did, however, stipulate that the Turkish government is to consult Iraq on matters concerning construction measures on the Euphrates, and the Ankara Protocol of 1946 provides for an exchange of hydrological data between Turkey and Iraq. See Scheumann 1996, p.44.

1987 provides for the establishment of a *Jordanian-Syrian Commission* with equal representation. If the commission is unable to reach agreement in disputes, it is required to report to the two governments, which are then expected to initiate negotiations aimed at resolving the dispute.[36] The Israeli-Jordanian Peace Treaty provides for the formation of *Joint Water Commission* with equal representation.[37] The 1960 Indus Agreement entailed the creation of a *Permanent Indus Commission* as well as an *Indus Basin Development Fund*, funded, among others, by the World Bank for the purpose of financing dams and irrigation canals in Pakistan and India. The *Permanent Indus Commission* has apparently continued with its work until today despite two Indian-Pakistani wars. The *International Commission for the Protection of the Rhine (ICPR)* was founded in 1950. Its tasks include the proposal of measures for the protection of the Rhine and preparatory work on possible agreements between the signatory governments.[38] It has, among other things, worked out recommendations for standards for emissions for individual pollutants; some of these re-commendations have been ratified by the signatories. The ICPR has no explicit monitoring function in connection with the Chloride Agreement.

2.8 Water Conflicts and Other Political Issues

Like all international agreements, the agreements examined here cannot be viewed independently of the general political relations between the signatory states. They are often tied into a complex of formal and informal agreements with reciprocal concessions, and water agreements are difficult to assess without knowledge of these.

The 1959 Nile Agreement, for instance, entailed an improvement of Sudan's position compared with the 1929 Nile Agreement. Sudan had refused to accept the 1929 agreement, arguing that it had been concluded by the colonial power, Britain, prior to Sudan's independence in 1956. For Egypt the agreement was important in that it ensured the country sufficient water for the Aswan Dam, where construction work started in January 1960, shortly after the agreement had been signed. The agreement implies recognition of Sudan's autonomy as a state, which had been disputed by Egypt a few years earlier under Gamal Abdel-Nasser, as well as by the 'Unionists' in Sudan with an eye to unifying the two states.[39] In the political constellation existing at that time the Nile Agreement thus meant a success for Sudan. Issues going beyond water are, however, not explicitly mentioned in the agreement.

This is different in the case of the 1987 Protocol of Economic Cooperation between Syria and Turkey. The agreement also deals with exploration for oil and gas deposits, the banking system, the transport of livestock, and customs formalities.

[36] Agreement 1987, Article 9.
[37] Peace Treaty 1994, Annex II, Article VII.
[38] Article 2 of the Bern Convention of 1963, quoted in: Durth 1996, pp.38-39.
[39] Warburg 1991, p.570.

Shortly thereafter the two states signed a security agreement requiring Syria to end its support of the Kurdish Workers' Party, PKK. Turkey accused Syria of not complying with the agreement, while Turkey claimed it had respected the provisions on the Euphrates' waters. Iraq and Syria are now demanding a greater share of the waters of the Euphrates. Turkey's willingness to make concessions will probably remain low as long as Turkey suspects Syria of direct or indirect support for the PKK.[40] In the long term, however, the fact that both parties could benefit from cooperation in areas beyond the utilization of water is working in favour of an agreement. An agreement not to support the PKK would be the foremost issue here. In addition, the planned networking of the electricity supplies between Turkey, Syria, Iraq, Jordan, and Egypt as well as the high-tension power line from the GAP area to Syria are creating conditions favourable to a possible agreement.[41] The networking project is set to be completed in 1998. However, the question of whether the benefits derived from this and other joint projects would, in the view of Syria and Iraq, compensate for the losses stemming from lower flow rates, is more than doubtful.

In the Israeli-Jordanian Peace Treaty of 1994, interlinkages with issues outside the area of water are obvious. It will not be possible here to attempt an overall assessment of this agreement,[42] which deals, among other things, with the initiation of diplomatic relations, border issues, economic cooperation, combating terrorism, and the status of Jerusalem. Even though the interlinkages between the individual areas during the negotiations are difficult for outside observers to grasp, it is likely that concessions on water issues will be linked with concessions in other areas.

2.9 Recommendations

Future international water agreements should consider the following aspects:

- International water agreements should comply with the customary-law principle of limited territorial sovereignty.
- Bilateral agreements on bodies of water with more than two riparian states should contain a reference stating that the signatory states seek *joint* negotiations with the other states concerned, and that water rights of other neighbouring states will be recognized as a matter of principle.
- Possible fluctuations in water flow rates should be taken into account. This can be ensured either by dividing the resources in accordance with a percentage-based key, or by dividing up a river system into individual rivers, whose flow-rate fluctuations are borne by the various signatory states.

[40] Beschorner 1992, pp.36-41; see also the interview with Süleyman Demirel in Der SPIEGEL, September 1996.

[41] See, e.g. Middle East Economic Digest, 6 September 1996, p.17. The power line is being funded on the Turkish side by the EU and on the Arab side by the Arab Fund for Economic and Social Development.

[42] See Schlünder and Ibrahim 1996.

- Issues concerning water quality and the protection of wetlands should be incorporated into international agreements, while respecting the polluter-pays principle.

Tab. 2.1. International water agreements in the Middle East

Designation	**Signed on**	**Signatories**	**Distribution of water resources (per annum)**	**Consideration of other neighbouring states**	**Distribution of the risk of fluctuating flow rates**
Agreement on the Full Utilization of the Nile Waters*	Nov. 1959	Egypt Sudan	55.5 BCM for Egypt 18.5 BCM for Sudan	Yes, but open	No arrangement
Protocol of Economic Cooperation**	July 1987	Turkey Syria	500 m^3/sec. at Turkish-Syrian border Rest for Turkey	Yes, but open	With Turkey alone
Agreement on the Utilization of the Waters of the Yarmuk River*	Sept. 1987	Syria Jordan	All tributaries rising on Syrian territory for Syria	No	Distributed but dividing the catchment basin
Agreement on the Sharing of the Euphrates	April 1990	Syria Iraq	58% for Iraq 42% for Syria	Implicitly, yes	Distributed by using proportional shares
Peace Treaty between Israel and Jordan*	October 1994	Israel Jordan	Jordan, except for 40 MCM for Israel Yarmuk, except for 45 MCM for Jordan	No	Israel (Jordan River) Jordan (Yarmuk River)

Source: Own compilation; * documented in the Annex, ** see Box 7.1., p.122

References

Agreement between the Hashemite Kingdom of Jordan and the Syrian Republic for the Utilization of the Waters of the Yarmuk River. Signed in Amman, 3 September 1987.

Beschorner, Natasha 1992: Water and Instability in the Middle East. Adelphi Paper 273, London, International Institute for Strategic Studies.

Biswas, Asit K. 1994: Management of International Water Resources: Some Recent Developments, in: Biswas, Asit K. (ed.), International Waters of the Middle East from Euphrates-Tigris to Nile. Bombay, Delhi, Calcutta, Madras. Oxford University Press.

BMZ (Bundesministerium für wirtschaftliche Zusammenarbeit und Entwicklung): Überlebensfrage Wasser - eine Ressource wird knapp, Entwicklungspolitik-Materialien Nr.94, Bonn 1995.

Chenevert, D. J. 1992: Application of the Draft Articles on the Non-navigational Uses of International Watercourses to the Water Disputes Involving the Nile and Jordan, in: Emory International Law Review, 6, pp.495-575.

Dellapenna, Joseph 1995: Building International Water Management Institutions: The Role of Treaties and Other Legal Arrangements, in: Allan, J. A., und Mallat, Chibli (eds.), Water in the Middle East: Legal, Political and Commercial Implications, pp.55-89. London, New York: I. B. Tauris.

Dombrowsky, Ines 1995: Wasserprobleme im Jordanbecken. Beiträge zur kommunalen und regionalen Planung, Nr.15, Frankfurt, Peter Lang Verlag.

Durth, Rainer 1996: Der Rhein - Ein langer Weg zum grenzüberschreitenden Umweltschutz, in: Aus Politik und Zeitgeschichte, pp.38-47.

Hofmann, T. (ed.): Wasser in Asien. Elementare Konflikte, Essen 1997.

Inan, Yüksel 1994: Legal Dimensions of International Watercourse (Euphrates and Tigris), in: Bagis, Ali Ihsan (ed.), Water as an Element of Cooperation and Development in the Middle East, pp.223-238, Ankara, Hacettepe University and Friedrich-Naumann-Foundation in Turkey.

International Law Association 1966: The Helsinki Rules.

Khassawneh, Awn 1995: The International Law Commission and Middle East Waters, in: Allan, J. A., und Mallat, Chibli (eds.), Water in the Middle East: Legal, Political and Commercial Implications, pp.21-28. London, New York: I. B. Tauris Publishers.

Kliot, Nurit 1994: Water Resources and Conflict in the Middle East, London and New York.

Lanz, Klaus 1996: Das Greenpeace-Buch vom Wasser.

Libiszewski, Stephan 1995: Water Disputes in the Jordan Basin Region and their Role in the Resolution of the Arab-Israeli Conflict. Zürich: ETH.

Lowi, Miriam R. 1993: Water and Power, The Politics of a Scarce Resource in the Jordan River Basin. Cambridge Middle East Library, 31.

McCaffrey, Stephen C. 1991: Seventh Report on the Law of the Non-navigational Uses of International Watercourses. United Nations.

McCaffrey, Stephen C. 1993: Water, Politics, and International Law, in: Gleick, Peter H. (ed.), Water in Crisis, pp.92-100. New York, Oxford, Oxford University Press.

Murakami, Masahiro 1995: Managing Water for Peace in the Middle East. Alternative Strategies, Tokyo, United Nations University Press.

Mutlu, Servet: The Southeastern Anatolia Project (GAP) of Turkey: Its Context, Objectives and Prospects, in: Orient 37 (1996)1, pp.59-86.

Peace Treaty between the State of Israel and the Hashemite Kingdom of Jordan, in: Internationale Politik, 7/1995, pp.86-108.

Scheumann, Waltina 1993: New Irrigation Schemes in Southeast Anatolia and in Northern Syria: More Competition and Conflict over the Euphrates?, in: Quarterly Journal of International Agriculture, Vol. 32, pp.240-259

Scheumann, Waltina 1996: Wasserkonflikt am Euphrat, in: INAMO, Nr.5-6, pp.38-45

Schlünder, Irene; Ibrahim, Ferhad 1996: Der jordanisch-israelische Friedensvertrag, in: Verfassung und Recht in Übersee 29(1), pp.73-94.
Shapland, Greg 1995: Policy Options for Downstream States in the Middle East, in: Allan, J. A., und Mallat, Chibli (eds.), Water in the Middle East: Legal, Political and Commercial Implications, pp.301-323. London, New York: I. B. Tauris Publishers.
United Arab Republic and Sudan: Agreement (with annexes) for the Full Utilization of the Nile Waters. Signed at Cairo, 8 November 1959, in: United Nations, Treaty Series 453, 1963, pp.63-77.
United Nations 1997: Convention on the Law of the Non-navigational Uses of International Watercourses, in: International Legal Materials, 3/1997, pp.700-720.
Warburg, G. R. 1991: The Nile in Egyptian-Sudanese Relations, in: Orient, 32 (4), pp.565-572.
World Health Organization (1989). Health Guidelines for the Use of Waste Water in Agriculture and Aquaculture. Technical Report Series 778.
World Resources Institute 1996: World Resources 1995-96, Washington D.C.

3 The Middle East Peace Process: Obstacles to Cooperation over Shared Waters[1]

Jochen Renger

Following the peace agreement of Camp David in 1979 there was great hope that economic relations between Israel and Egypt would contribute to a peaceful settlement of the Arab-Israeli conflict. However, this hope fostered by the international community remained unfulfilled. The agreement of Camp David remained a cold peace. In the current peace process, which began in Madrid in 1991, the Western countries, some Arab states and various international organizations favour regional economic cooperation as the central means for establishing a new order in the Middle East. The conflicting parties should cooperate within the fields of economics, security, environment, refugees and water. Cooperation is supposed to resolve their common problems and thus lead to the integration of the entire Middle East. In the end, a new era in the Middle East is to be expected, as outlined by Shimon Peres, the former Israeli Prime Minister, or the World Bank.[2]

This article addresses the obstacles as well as the perspectives of cooperation among the countries that border the Jordan River in the field of water resources. The current peace process has already produced some results in resolving the water disputes in a cooperative way. However, it is at a standstill. The main reason is that the conflicting parties have different objectives for the individual benefits of the future peace order. Israel, in particular, seems to be immovable because it has realized its objectives to a substantial degree already.

The peace treaty between Israel and Jordan signed on October 26, 1994 laid down the use of water resources of the Jordan and the Yarmuk rivers, as well as the use of ground water resources in the Arava Valley.[3] Israel granted the Palestinians an increased quota of water for municipal use and limited rights to administer these water supplies.[4] Within the framework of the multilateral working group on water, some projects for the improvement of water use were initiated.[5] However, there is no reason for optimism as far as the cooperative water policy in the Jordan River Basin is concerned. The conflicting parties are still far away from a reasonable and equitable distribution of the water resources according to interna-

---

[1] I am greatly indebted to Matthias Ruchser and Andrew Kandel for their invaluable support.

[2] Peres 1993; World Bank 1992.

[3] Peace treaty between the State of Israel and the Hashemite Kingdom of Jordan, Annex II, October 26, 1994.

[4] Libiszewski 1995, pp.77-81.

[5] Renger 1995.

tional standards. The common management of the shared water resources in the basin looms out of sight. The goal for future water cooperation should be a regional water regime on the basis of international law, that settles the disputes on water rights, water allocation and water use. Regional experts agree that the water supply of the Middle East will only be secured by multilateral cooperation. This will be one condition for a long-lasting peace in the region.

The problem of water overlaps three areas of conflict in international politics: power, security and welfare.[6] Power refers to the distribution of immaterial possibilities, such as political participation or equity. Security contains the protection of the physical existence against threats. Welfare describes the allocation of material chances of survival. The water issue cannot be analyzed separately from the various questions of the complex Arab-Israeli conflict. Since the parties to the conflict will benefit differently from a new regional order, the varying positions of goal attainment among the parties has led to a standstill of the negotiations.

3.1 Power over Territory and Water

From a strategic point of view the controversial territories in the Arab-Israeli conflict, i.e the Golan Heights, the West Bank and South Lebanon, are extremely important. Two of the three tributaries to the Jordan River originate in the Golan Heights area. The West Bank contains rich ground water resources. The peace negotiations up to now have shown the complexity of the water problem: Questions of water are always those of power. The occupation of the Golan Heights, the West Bank and South Lebanon enables Israel to control unilaterally the water resources of the region. At present, Israel is meeting half of its water demand with water resources located outside its internationally recognized borders.[7] Thus, the disputed territories mean physical access to water and its use. If the formula of the current peace process 'land for peace' were realized, Israel would have to give up control of its acquired water resources. From the Israeli perspective, that would mean losing water.[8]

The detailed arrangements between Israel and Jordan on water allocation of the Jordan and Yarmuk rivers were not founded on common economic and security interests, but rather because the two countries had no territorial disputes. The water conflict between Israel and Jordan was a 'genuine water dispute'.[9]

Within the framework of the peace agreement Israel formally entitled to the Palestinians water rights. However, this does not imply Palestinian sovereignty over water resources. The Palestinian Authority was only granted a few roles in water administration, such as in water supply and waste water management. Ac-

---

6 The approach of using power, security and welfare as a higher order of international politics is based on Czempiel (1981).

7 Stauffer 1996, p.9.

8 Announcement on the significance of land and water presented by the Israeli Ministry of Agriculture, The Jerusalem Post, August 10, 1990.

9 Libiszewski 1995, p.36.

cording to the Oslo-II Agreement, a bilateral commission is responsible for all the decisions that are being taken in the water sector. All decisions in the commission have to be reached unanimously. This reduces the Palestinian significance in the water issue to a minimum. Despite the fact that Israel and the Palestinian Authority are working together in some areas of water management, this does not mean they are cooperating. The water is being supplied by Israel, which dictates the quota and the prices. The issue of water between Israel and the Palestinians is directly linked to the scope of Palestinian sovereignty. Sovereignty over territory means control of water resources. The reconstruction of the Palestinian community, and, in particular, its economy, depends on water. Not until the final status of the Palestinian territories has been determined will the redistribution of water resources be resolved.

Jordan will also be affected by the future status of the Palestinian territories. It must share the Lower Jordan River with the Palestinians. Therefore, it is no surprise that Jordan and Israel were able to settle their water disputes rather quickly. Jordan wanted to create facts of the matter in order to forestall Palestinian claims.

Syria has made it known that without the settlement of the territorial dispute, there will be no talks on any other issues. The Syrian government insists on the unconditional Israeli retreat from the annexed Golan Heights according to the various UN-resolutions such as 242. With the exception of the local people, the waters from the Golan Heights do not substanially contribute to the Syrian water budget. For the Syrian leadership, however, the Golan Heights are an essential question of national prestige. This is very important for President Asad to legitimize his power. The Golan Heights are ingrained as a symbol of national dignity. Within the Arab world, a heroic steadfastness is being conjured up. Leaving out the territorial question in favour of a pragmatic approach to the water issue would be inconceivable to Syria.

The former Israeli Prime Minister Peres and the late Prime Minister Rabin linked the Golan Heights mainly to strategic security considerations. It seemed possible that an extensive security regime under the auspices of international forces might resolve the conflict on the Golan Heights. Due to ideological reasons, the new government under Prime Minister Netanyahu refuses to put the Golan Heights on the agenda of the peace negotiations.

In consideration of an Israeli-Lebanese peace treaty, only Syria stands in the way of further negotiations. Israel has no territorial claims towards Lebanon but demands security on its northern frontier. The Lebanese government is assumed to be capable of controlling the Southern Lebanese territory politically and militarily. For both states an agreement would be very desirable. Lebanon needs peace to reconstruct its economy, Israel needs peace due to security aspects. However, the Lebanese government is adjutant to Damascus, making no solo run possible.

3.2 Security and Water

Water is being considered by the conflicting parties as a strategic resource. To Israel, water is a question of existence for the Jewish state. Half of the Israeli water supply depends on water resources which are located outside its internationally recognized borders. In the past Israel has been confronted with the experience that water was used as a weapon by Arab countries. In 1964 the Arab states, under the leadership of Egypt and Syria, tried to divert the tributaries of the Jordan River (Arab Diversion Plan). In gaining control of the Jordan River springs during the Six Day War of 1967, Israel eliminated this threat. These events demonstrate that the water problem of the Jordan Basin has a security dimension.[10] Today, the possibilty of fighting over water supplies seems marginal, but it is likely that water might be used as leverage during a conflict. To Israel, the question of water in the current peace process is particularly linked to its security. Israel refuses to give up a substantial share of the water resources in the occupied territories either to the Palestinians, Syria or to Lebanon.[11]

For the Palestinians, water is also a part of their strategic security needs. Israel used water to implement its power interests in the West Bank and the Gaza Strip.[12] The control over Palestinian water resources has enabled Israel to hold down the socio-economic development of the Palestinian territories. Sovereignty over the West Bank's water resources is an essential part of the Palestinian's security conception. The question of water is one of Palestinian sovereignty and independence.

3.3 Welfare and Water

The availability of water determines the socio-economic development of the states of the Middle East.[13] With the exeption of Lebanon, all parties to the conflict suffer from water scarcity. The concept of the Water Stress Index states that the annual minimum water requirement for life is 500 m^3 per capita. Israel (300 m^3), the Palestinian territories (165 m^3) and Jordan (300 m^3) do not even approach this minimum requirement.[14] Water use greatly exceeds water supply and can only be satisfied by the acceptance of serious ecological damage. In this decade Jordan and Israel have met 20% of their water needs by exploiting non-renewable ground water resources. Despite the arid climate of the Middle East, the water crisis has been exacerbated by the fact that water is used especially for agriculture based on irrigation. The 'thirsty' agricultural sectors consume from 75% (Israel and Jordan) to 90% (Syria) of the national water resources without contributing substantially to

[10] Schmid 1993, pp.102-104.
[11] See Schiff (1993, pp.90-93) for the Israeli-Syrian context.
[12] Renger and Thiele 1996, pp.78-80.
[13] Gleick 1993, pp.89-93.
[14] Shuval 1992, p.134.

the economy as far as the GDP or the employment rate is concerned.[15] None of the parties to the conflict will be able to afford to renounce water supplies in favour of a regional re-allocation of water resources, specifically due to various ambitious development plans in the Middle Eastern agricultural sector.

Considering the present negative Israeli water balance, the question is where the water should come from in case of a re-allocation of water resources within the framework of a future peace agreement. The Israeli government argues that no concessions in water politics could be made due to the scarcity of water. There are technical options available to increase the water supply, such as sea water desalination or water imports, but these options would be very expensive and therefore only possible through foreign aid[16]. Rationing the use of water in the agricultural sector would be one of the strategies in providing water for other uses. However, due to political and ideological reasons, no Israeli government would consider the option of a structural change in the economy. Other solutions to resolve this water dilemma are being discussed. Israel could return the occupied territories along with the water resources and would have to purchase water from Syria instead. Taking the Israeli financial situation into account, this is not an option either.

The Palestinians suffer most from the asymmetric water distribution in the Jordan River Basin.[17] Not only do they receive the lowest amount of water per capita, they also do not have any sovereignty over water resources. The Palestinians depend solely on Israeli water rations.[18] Due to the fact that the West Bank borders the Jordan River, the Palestinians claim a share of the Lower Jordan River waters. However, Israel stores most of the Jordan River waters in Lake Tiberias and pumps it through the National Water Carrier to the coastal area and the Negev. The result is that only a rivulet reaches the Palestinian territory. The Palestinians continue to insist on a higher water quota from the Israeli occupation force as well as on their water rights according to international law. Water is being considered as a means of reconstructing Palestinian society in the occupied territories, particularly in the economic sphere. The Oslo-II Agreement has brought partial results. The Palestinians were granted a higher, but still marginal, water quota.

Because of the ambitious Turkish plans to store the waters of the Euphrates River, Syria has been forced to mobilize its various water resources in all parts of the country. Particular attention is directed towards the Upper Yarmuk River which is considered solely as a Syrian river. As a result of the Syrian construction of dams, the water flow towards Jordan and Israel will be reduced. This is of particular concern to Jordan, which guaranteed water quotas to Israel in the peace treaty of 1994. It is possible that Jordan will lose a considerable amount of water. Unless a multilateral water agreement between the countries bordering the Jordan River Basin is made, Jordan will have to make new arrangements with Syria and Israel.

---

[15] Renger and Thiele 1996, p.76.

[16] Stauffer 1996, pp.12-17.

[17] United Nations 1992.

[18] Lowi 1993, pp.117-122.

3.4 The Present Standstill of the Peace Process

As the discussion above has demonstrated, the complexity of the water problems lies with the interconnection of the various dimensions of the Arab-Israeli conflict. The chances of cooperating in the water issue are directly linked to the future peace process. The euphoria which resulted from the progress achieved between Israel and Jordan on the one hand and Israel and the Palestinians on the other hand can only mask temporarily the fact that deep differences of opinion do exist. Presently, the positions of the conflicting parties seem to be immovable, the main reason being that the parties have different objectives for the individual benefits of a future peace order in the region.[19]

Israel's main objectives are to achieve security and welfare. These objectives have been realized to a substantial part already. The state of Israel has been recognized by the Palestine Liberation Organization (PLO) and is no longer called into question by many Arab states, despite the fact that Israel has not yet returned the occupied Arab territories. The Arab doctrine to boycott Israel due to the occupation of Arab territory no longer exists. Official political relations and contacts between Israel and various Arab States, with the exception of Egypt and Jordan, are still delicate. However, in diplomatic and especially economic circles, the parties are beginning to arrange their affairs. The Israeli military operation in Southern Lebanon in 1996 has demonstrated that Israel is able to find a security arrangement with Lebanon, although the Arab-Israeli conflict has not yet been resolved. Israel has been able to restrict the conflict with the Hizbollah to within the occupied security zone. From an economic point of view, Israel has already benefited from its 'concessions' made in the various peace treaties. The United States granted financial aid in return for recognizing the PLO. In 1995 the European Union offered Israel a new association agreement including privileges for Israeli agro-exports to Europe. As a result of the abandonment of the secondary and tertiary Arab boycott against the Jewish state, Israel has gained a foothold in the global economy, in particular in Asia. Because Israel has already reached many of its objectives in the current peace process, there is no urgent need for Israel to make further concessions, especially in the very sensitive issue of water.

In terms of further negotiations, the Palestinian Authority is in a very weak position. The PLO was obliged to join the peace process in order to save the organization from political insignificance and financial ruin.[20] The independent state of Palestine as the main goal of the PLO had to be postponed for an indefinite period. In order to achieve the Oslo-I and the Oslo-II agreements, the Palestinian leadership gave up its most important trump cards, e.g. recognition of the state of Israel and non-aggression. The Palestinians' only hope is that Israel will make further concessions in the future peace talks. However, without pressure from the United States on Israel, the Palestinian Authority is helpless and powerless. Even the implementation of the existing treaty is a dogged undertaking as the Hebron

[19] Perthes 1995.

[20] Beck 1996, pp.9-11.

agreement has recently shown. The Gaza Strip and the Jericho area are hydrological crises areas which cannot be saved without international financial aid and technical assistance. As far as water supply is concerned, the Palestinians will continue to depend on Israel.

From the Syrian regime's point of view a future peace order has to be considered as a risk which cannot be calculated.[21] The benefits of a peace agreement remain completely unclear to President Asad. The current 'front position' towards Israel is a very comfortable one since it increases the political role of Syria in the region and the world. Despite the fact that this role exceeds its military and economic potential, Syria uses it to receive financial revenues from various Arab and European countries. Syria had the Western countries pay for its moderate involvement in the Second Gulf War of 1990/91. Economically Syria is ill prepared for a new peace order in the region, which will be based on trade and economic cooperation. The ailing socialist economy dominated by the state cannot compete with other economies in the region. Syria's perspective of a regional, economic integration lies with Iraq and Iran, but these states are excluded from the peace talks. Thus, the Syrian regime needs time to adjust its economy in order to play a more serious role in the peace process.

Lebanon is preparing itself for peace: its future economic role in the region will be that of a financial centre, service centre and tax oasis. But without Syrian permission, no peace with Israel will be achieved.

Jordan and Israel signed a peace treaty in 1994. The Hashemite leadership hoped to resolve its financial crisis by making peace with Israel.[22] King Hussein also tried to protect his religious claim on the holy places of East Jerusalem. In addition, the Jordanian water crisis forced an arrangement with Israel on the Yarmuk River in order to increase Jordan's water supply. However, Jordan's hopes were disappointed: the financial aid by the Western countries was used to pay off debts; development aid is being conditionalized, and the expected benefits from trade have failed to materialize. At least the water disputes could be settled. Israel and Jordan fixed quotas of water allocation and established a bilateral water regime (water exchange, water commission), although other riparians were excluded.[23] In the event that a greater regional water cooperation is established, a new arrangement will have to be found.

[21] Perthes 1995, p.58.

[22] Due to the ambivalent policy of King Hussein in the Second Gulf War of 1990/91, Saudi Arabia cut off financial subsidies to Jordan. The Hashemite leadership has been trying to mobilize financial aid from Western countries by supporting the peace process.

[23] Peace treaty between the Hashemite Kingdom of Jordan and Israel, Annex II, October 26, 1994.

3.5 Outlook

In the foreseeable future there is no prospect of water cooperation between the bordering countries of the Jordan River Basin because the water issue is instrinsically linked to the central issues of the Arab-Israeli conflict. A regional water agreement will depend on the resolution of the entire conflict. Questions of security and territory have precedence over sheer questions of water. At the same time, increasing water scarcity restricts the scope of re-allocating the regional water resources. Israel, which has monopolized most of the water resources, currently has no interest in giving back a substantial water share to its Arab neighbours. Nevertheless, this does not mean that the situation is a lost cause. There is still room for dynamic development in the water issue. First, it would be possible to re-allocate water resources within a package solution. Due to the fact that the tributaries of the Jordan River play a small role for the Lebanese and Syrian national water supplies, a water-for-concessions-solution could be taken into consideration. Israel would have to pay for Lebanese and Syrian water on a financial or material basis (goods). Such a solution would include international guarantees to secure the flow of the Jordan River feeders to Israel. Second, there is a high potential in all countries of the region to conserve water by optimizing the efficient use of water. This would increase the opportunity to re-allocate the regional water resources, in particular between Israel and the Palestinians. The international community would have to help to improve the water balances. Water imports as well as desalination plants are already being projected. It would be desirable to promote efficient use of the water according to the needs of the national economies. This means initiating an economic structural change, so as to reduce agriculture based on irrigation. Third, the parties to the conflict share common interests as far as the ecological protection of the Jordan River basin is concerned. No party wants the ecosystem to be destroyed, which would result in pollution or hypersalination of water. Such interests could give an impetus to the peace talks on water. The common ecological interests could function as the focal point for further water cooperation. However, the present status of the Middle East peace process does not leave room for optimism.

References

A Strategy for Managing Water in the Middle East and North Africa 1994, Washington, D.C.

Beck, Martin 1996: Der israelisch-palästinensische Friedensprozeß, in: Aus Politik und Zeitgeschichte, No.18, pp.3-12.

Czempiel, Ernst-Otto 1981: Internationale Politik. Ein Konfliktmodell, Paderborn/München/Wien.

Dombrowsky, Ines 1995: Wasserprobleme im Jordanbecken. Perspektiven einer gerechten und nachhaltigen Nutzung internationaler Ressourcen, Frankfurt a.M., Bern.

From Scarcity to Security. Averting a Water Crisis in the Middle East and North Africa, 1995, Washington, D.C.

Gleick, Peter H. 1993: Water and Conflict. Fresh Water Resources and International Security, in: International Security, Vol.18, No.1, pp.79-112.
Halbach, Axel J. 1995: Kooperation im Nahen Osten. Halten sich Hoffnung und Ernüchterung noch die Waage?, in: IFO-Schnelldienst, Vol.48, No.9, pp.14-20.
Libiszewski, Stephan 1995: Water Disputes in the Jordan River Region and their Role in the Resolution of the Arab-Israeli Conflict (ENCOP Occasional Paper No.13), Zürich.
Libiszewski, Stephan; Schiffler, Manuel (eds.) 1995: Wasserkonflikte und Wassermanagement im Jordanbecken. Beiträge eines Colloquium in Berlin, 16.-17. Juni 1995, Berlin.
Lowi, Miriam R. 1993: Bridging the Divide. Transboundary Resource Disputes and the Case of West Bank Water, in: International Security, Vol.18, No.1, pp.113-138.
Middle East Water Commission 1995: Observations Regarding Water Sharing and Management. An Intensive Analysis of the Jordan River Basin with Reference to Long-Distance Transfers, in: Water Resources Development, Vol.11, No.4, pp.351-375.
Pawelka, Peter 1993: Der Vordere Orient und die internationale Politik, Stuttgart, Berlin, Köln.
Peres, Shimon 1993: Die Versöhnung. Der neue Nahe Osten, Berlin.
Perthes, Volker 1995: Integration oder Trennung? Die Logiken des nahöstlichen Friedensprozesses, in: Internationale Politik, Vol.50, No.7, pp.55-60.
Renger, Jochen 1995: Die multilateralen Friedensverhandlungen der Arbeitsgruppe Wasser, in: Asien Afrika Lateinamerika, Vol.23, No.2, pp.149-157.
Renger, Jochen; Thiele, Andreas 1996: Politische Verteilungskonflikte um Wasserressourcen. Wassernutzung und Wasserverteilung im Jordanbecken. Israel und seine arabischen Nachbarn, in: Der Bürger im Staat, Vol.46, No.1, pp.74-82.
Schiff, Ze'ev 1993: Peace with Security. Israel's Minimal Security Requirements in Negotiations with Israel (The Washington Institute Policy Papers No.34), Washington, D.C.
Schiffler, Manuel 1995: Das Wasser im Nahostfriedensprozeß. Ansätze zu einer gerechten Aufteilung und Möglichkeiten zur Entschärfung der Wasserknappheit, in: Orient, Vol.35, No.4, pp.603-624.
Schmid, Claudia 1993: Der israelisch-arabische Konflikt und die Bedeutung des Vorderen Orients als sicherheitspolitische Region nach dem Ost-West-Konflikt, Baden-Baden.
Shuval, Hillel 1992: Approaches to Resolving the Water Conflicts Between Israel and Her Neighbours. A Regional Water-For-Peace-Plan, in: Water International, Vol.17, No.3, pp.133-143.
Stauffer, Thomas R. 1996: Water and War in the Middle East. The Hydraulic Parameters of Conflict (The Center for Policy Analysis on Palestine, Information Paper No.5), Washington, D.C.
Syrien. Frieden mit begrenzter Wirkung, in: Der Überblick, 1995, Vol.31, No.4, pp.56-60.
United Nations 1992: Water Resources of the Occupied Palestinian Territory, New York.
Wolf, Aaron T. 1995: Hydropolitics Along the Jordan River. Scarce Water and its Impact on the Arab Israeli Conflict, Tokyo, New York, Paris.
Wolff, Peter 1992: Durstiges Israel. Eine kritische Betrachtung zur wasserwirtschaftlichen Situation Israels (Gesamthochschule Kassel Fachbereich Internationale Agrarwirtschaft, Arbeiten und Berichte Nr.28), Kassel.
World Bank 1993: Developing the Occupied Territories. An Investment in Peace, 6 Vol., Washington, D.C.

4 Transboundary Externalities and Regional Integration

Rainer Durth

Due to population growth and higher per capita use, human water use has increased tenfold over the last hundred years and is approaching 80% of the volume of fresh water that currently can be harnessed on a basis, both economically and ecologically sustainable.[1] Since most of the fresh water supply is surface water from rivers and lakes, international river systems, covering almost half the land surface of the earth and being home to 44% of the world's population,[2] will become an increasingly contentious issue over the next few decades. Unless the competing claims of the riparian states can be satisfied through cooperation, they will lead to international conflicts in the medium term, as the current tensions in the Middle East exemplify. Upstream-downstream problems are problems of externalities in which the external effect always operates in the same direction, i.e. downstream. The classical economic solution to external effects is internalization: only if all parties adjacent to the river have to bear the full economic consequences of their actions will optimum utilization of the river be achieved.[3]

Both demographic and economic development began much earlier in Europe than elsewhere. Furthermore, since World War II the European Community has developed a particular method for dealing with environmental problems on international rivers. This article will therefore compare legal and political conditions for cooperation on international rivers in non-integrated regions (section 1) with the political and legal framework for cooperation inside the European Union (section 2). Empirical evidence on the cooperation process in the Rhine River Basin and the Euphrates River Basin (section 3) will complement the theoretical insights. Finally, the European experience will be summed up in ten rules for solving upstream-downstream conflicts over international rivers (section 4).

---

[1] Meadows et al. 1992, p.79.
[2] Biswas 1993.
[3] Coase 1969.

4.1 International Upstream-Downstream Problems in Non--integrated Areas

Any analysis of upstream-downstream problems in non-integrated areas must focus on the beginning of cooperation. In non-integrated regions of the Third World in particular, it is extremely difficult to initiate cooperation between upstream and downstream states. We shall therefore first discuss conditions necessary for cooperation between states (section 1.1). In order to analyze the governments' perspective on cross-border negotiations, we will then turn to the rules of international law, under which governments are the key actors for any solution (section 1.2).

4.1.1 Necessary Conditions for Cooperative Solutions

Solutions for international upstream-downstream problems can only be developed if cross-border cooperation can be initiated. Cooperation will come about only if the riparian states are prepared to submit themselves to credible sanctions in the event of infringements of treaty obligations.[4] A simple way of meeting this requirement is to 'exchange hostages' or to create mutual dependency. Applying this principle, European integration after World War II began with the creation of the European Coal and Steel Community in 1952, in which member states were to merge their industries of strategic war-time importance. Without such security guarantees, even major efficiency gains may have been unattainable as the water conflicts in the Middle East show.[5]

Secondly, empirical analysis shows that, despite large potential efficiency gains, there will be no cooperation if upstream and downstream states have radically different ideas about the fair usage or division of international resources.[6] For this reason all parties must first concentrate on diminishing the differences between their perceptions of what is just and equitable. An important initial step in such a process may consist of distinguishing between a *just outcome* and a *just procedure* for resolving conflicts of interest over water usage. All in all, it is easier to reconcile differing conceptions of *just procedures*. Furthermore, implementing economically efficient rules should be encouraged, since this will increase efficiency gains and, hence, augment the potential for compensatory payments. Since negotiations may be broken off if differences in negotiation power are too big, it might be advantageous for a powerful party to deliberately weaken its own position in order to ensure a peaceful settlement. This can be achieved, for instance, by eliminating possible threats and by influencing risk perceptions and time preferences of the negotiating parties.

---

4 E.g. Barrett 1992; Mäler 1991; Ströbele 1991; Kuhl 1987.
5 See section 3 of this paper.
6 Durth 1995.

4.1.2 The Decisiveness of Governments' Perspectives

International law on the environmentally sound management of international rivers is only recently developing.[7] There are attempts to regulate relations between sovereign states represented by their governments. As a rule, international law provides no basis for individual claims and enables governments to monopolize international negotiations over transboundary environmental problems.[8] Apart from international agreements, which may be quite diverse, three major principles of international environmental law have been established up to the present:

- Substantial and unusual pollution originating in one country and resulting in damaging or endangering effects on the territory of another country must be prevented.[9]
- Both countries, upstream and downstream, have a duty to cooperate.[10]
- In case of an emergency, downstream countries should have access to all relevant information held by the ones upstream.[11]

International environmental law knows neither applicable emission nor immission norms, nor are there enforcing mechanisms or institutions. Every time a state is called to the International Court of Justice it has to consent to the trial. In addition, the burden of proof lies entirely with the pollutee.

Under international law, national governments are in total control of the internalization process. On the international level the pollution problem will be solved (or not) through a bilateral monopoly: both governments bargain for a maximum share of the possible efficiency gains. On the national level the governments have monopoly positions too, concerning all transboundary negotiations as well as all transboundary flows of information relating either to international negotiations or to the pollution problem itself. This is a rather comfortable position for the governments concerned, because it enables them to cash in a monopoly rent on solutions for transboundary pollution problems. Given that governments pursue individual goals rather than national welfare, naturally they will sell the results of international negotiations to the domestic (or foreign) interest group which grants them the largest (political or economic) profit in return. Decisions on international problems are even more attractive for a government than decisions on domestic matters if the government can be sure that its opportunistic behaviour will not be monitored by the electorate. An additional monopoly on transboundary informa-

---

7 See Barandat and Kaplan in this volume. For a wider survey, see Kunig 1992; Kiss and Shelton 1991; specifically for international watercourses, see Lammers 1989.

8 Individual claims are subject to national law. The rules of international private law regulate whether domestic or foreign law should be applied. Until now, transboundary individual claims for environmental damage are rather rare. See O'Connell 1992, pp.294; Caponera 1987, pp.17; Rest 1986, pp17.

9 Principle 24 of the Stockholm Declaration; see also Kiss and Shelton 1991, p.131; Luge 1989, p.62.

10 Principle 24 of the Stockholm Declaration; see also Kiss and Shelton 1991, p.131.

11 Principle 9(1) UNEP Shared Resources Principles; see also Kiss and Shelton 1991, p.135.

tion - as it is provided by international law - will, thus, be welcomed enthusiastically by the governments, because it enables them to pocket maximum rents without paying political costs. Any unpopular results of international negotiations can doubtlessly be blamed on foreign governments.

International cooperation in non-integrated regions will come about only if all the governments involved expect cooperation to give them more advantages than disadvantages. The cost-benefit-analysis conducted by governments will be influenced primarily by the following factors:

- *Improvements in efficiency,* being the (economic) driving force behind cross-border cooperation.
- *Compensation* or simultaneous settling of several international problems.
- *Effects on the discretionary freedom of action* of the governments concerned.
- *Effects on the political prestige* of the governments involved.

Although compensation is part of the classical economic solution for upstream-downstream problems, they play a very subordinate role in reality. Much more important for explaining and solving conflicts on international rivers are effects on the prestige and popularity of the governments involved.

4.2 European Integration Has Changed the Internalization Framework

For centuries, European riparian states had to cope with the problems described above. After World War II, however, cooperation was facilitated significantly, and solutions for problems on transboundary externalities have been realized in great numbers. In order to understand the effects of European integration we shall, again, focus on the conditions for interstatal cooperation. Thereafter, the legal framework inside the European Union will serve as a basis for analyzing the perspective of governments and private parties involved in the externality problem.

4.2.1 International Cooperation Inside the European Union

Regarding riparian states as homogeneous blocks, international cooperation between the member states of the European Union seems much easier than cooperation between the states in the Middle East. Since the European states have been cooperating on a wide array of issues over a long period, it is much easier for them to make binding commitments: any default will damage their reputation and reduce the willingness of the partner states to cooperate on parallel or future issues. Furthermore, mechanisms of settlement are more elaborated, facilitating transfer payments, focusing on *just procedures* (not the outcomes themselves) and ensuring a good understanding of the partners' situation.

4.2.2 Political Economy of Governmental and Private Action in the European Union

The legal framework for transboundary river pollution inside the European Union is much more complex than under international law. Complementary to the rules of international law and to national rules, four important levels of community law may be differentiated.[12]

- Water quality protection. Community law includes a long list of directives with precise emission and immission norms. Emission norms are applicable for specific dangerous substances, e.g. heavy metals. Immission norms are relevant if the water is used for a specific purpose, e.g. fishery, bathing or drinking.
- Protection of other environmental media. The regulation of air pollution, dangerous waste, fertilizers or endangered species has indirect but substantial positive effects on the quality of European water resources.
- Horizontal measures. Medial environmental policy is supplemented by so-called horizontal measures. Most of these horizontal measures are intended to improve the exchange of environmentally relevant data and make them readily accessible to all EU-citizens. As the European Commission did not have the means to control the effective application of their directives in the member states, it has significantly improved the conditions for public and decentralized control of the directives.[13]
- Integration of European environmental policy into other policies of the Union. The formulation of all other policies of the EU, e.g. regional development, energy or transport, must take into account environmental concerns.[14]

In contrast to international law, European law contains an elaborate set of emission and immission norms. In addition, it possesses more effective enforcement mechanisms and institutions, even if the polluter does not consent to a trial. During the last 15 years the burden of proof has been shifted systematically from the pollutee to the polluter.

The changes on national political markets which have resulted from the horizontal measures described above are particularly interesting for the internalization of international externalities: While under international law, transboundary environmental problems are a matter of intergovernmental information and negotiations, allowing these problems to be tackled directly by interest groups in Europe. These

---

[12] For a detailed and systematic description of European environmental law, see Johnson and Corcelle 1992; Paleologou 1993.

[13] Horizontal directives provide, i.a. free access for all EU-citizens to all environmental information being available in any administration in any member state; a duty to realize environmental impact studies for projects which may have significant impact on the environment; duties of the member states to report publicly on environmental achievements; the establishment of a European Environmental Agency; eco-labels for products and eco-audits for enterprises; a directive for environmental liability is under discussion.

[14] EU-Treaty, Art. 130 r 2, 2.

can now establish additional transboundary information flows, collecting information abroad and distributing it at home or collecting/generating information at home and distributing it abroad. International information flows enable interest groups to control intergovernmental negotiations and to influence political processes at the same time. Furthermore, interest groups can more easily participate in foreign administrative or legal procedures. Participation depends on the procedure in question. Some EU-directives grant direct rights concerning foreign administrations, such as the directive on free access to environmental information. Finally, European interest groups have the option of negotiating directly with foreign interest groups. Such direct interest group negotiations may take place between affiliated groups, e.g. ecologists forming an international alliance, or between rival groups looking for a 'peaceful' settlement, e.g. chemical industries avoiding highly publicized trials with water works.

The legal setting for transboundary pollution problems in integrated river basins results in a fundamental change of governmental freedom compared to the situation in non-integrated areas: both government monopolies, i.e. transboundary information flows and international negotiations, have been diminished: first, governments are no longer the only provider of international negotiations on domestic 'political markets'; on the contrary, their transboundary activities compete with transboundary activities of other private and administrative actors. As a consequence of this competition the political price for intergovernmental negotiations is likely to fall. Secondly, while governments control transboundary information flows in non-integrated areas, transboundary information is free for everyone in the European Union. Consequently, it becomes easier for domestic interest groups and the electorate to monitor governmental activities. This reduction of informational asymmetries, too, will reduce the price of intergovernmental negotiations on political markets. Being a function of the falling price, the political demand for intergovernmental activities should rise between integrated states.[15] An empirical analysis of all international agreements on transboundary rivers in Europe and North America confirms this expectation.[16] The number of international treaties on transboundary rivers increased dramatically as soon as governments had lost their respective monopolies in the course of European integration.

While governments forfeit a lot of their relative power and might even be regarded as the losers of integration, private interest groups are surely among the winners. Instead of relying solely on the promises of their national governments to solve the problem for them, they get a realistic chance to influence the political process abroad and at home, as well as to start negotiations on their own. The action set of private interest groups which want to influence the internalization of transboundary externalities is, thus, significantly greater in integrated areas. In the European Union private interest groups who feel (economically) concerned by transboundary pollution can choose between four alternative ways to start/ influence the internalization process:

---

[15] At least as long as governments have comparative advantages in international negotiations.

[16] Durth 1996d.

- They may delegate internalization to the national government;
- They may engage in transboundary information flows,[17]
- They may participate in foreign administrative and/or legal acts, and
- They may negotiate directly with other interest groups.

Rational interest groups will develop an optimal combination of all four alternatives. This optimal mix will differ from group to group, depending on the characteristics of every interest group, e.g. the conditions for collective action of the members[18] or the particularities of the topic concerned.[19]

This transfer of power from governments to private parties is efficient from a polit-economical point of view for two reasons: first, it increases the competition among supply and demand in national political markets, reducing governmental discretionary power, and consequently reducing the potential for rents. Secondly, it delegates decisions on transboundary environmental problems to those people who are most concerned by them and who are most likely to possess the maximum information required.

The conditions for transboundary internalization in Europe are by no means given, nor are they evident; rather, they are the result of a long and open-ended learning process. Contrary to popular belief, the internalization of international environmental externalities is rather easy in Europe because it is less based on mighty Eurocrats in Brussels, depending more on economic (or other) interests of individuals. The European Commission has limited its involvement in setting up the framework for an efficient internalization process. This framework is aimed at weakening the two following important asymmetries in national political markets:

- Traditionally, domestic interest groups have a home advantage in the domestic political market for distorting efficient results;
- Furthermore, interest groups are very different with regard to their organisational capacities and political influence. Once an externality exists, the supporters of the externality tend to be stronger than their opponents.[20]

Internalization will be made easier if political markets are transformed: foreign interest groups must not be discriminated against and the position of supporters of internalization must be strengthened systematically. Environmental law in the European Union has already taken some steps in this direction, but some significant asymmetries prevail.[21] Nevertheless, from a legal perspective the development of concepts for the solution of international environmental problems in Europe is less a question of intergovernmental bargaining, rather, it takes place between the individuals (or groups of individuals) concerned. This is a very important development in the search for transboundary internalisation schemes.

---

[17] Influencing and controlling internalization activities at the same time.

[18] Olson 1991; Sandler 1992.

[19] E.g. the potential for directly influencing political processes may be different for environmental and industrial purposes.

[20] E.g. environmentalists vs. industrialists.

[21] E.g. polluting industries could be obliged to publicize all environmentally relevant information; burden of proof could be transferred from supporters to opponents of internalization; an effective environmental liability scheme could be implemented.

The following Table 4.1. summarizes the 'typical' features of upstream-downstream problems in non-integrated regions and in the European Union:

Table 4.1. Upstream-downstream problems in non-integrated regions and the European Union

Upstream-downstream problems in non-integrated regions and the EU	**Non-integrated regions**	**European Union**
1. Requirements for international cooperation		
• Binding commitments	very difficult	easier
• Convergence of ideas of a fair utilization	very difficult	easier
• Negotiation powers of the riparian states	rather unequal	more balanced
2. Role of governments		
• Monopoly on transboundary negotiations	rather strong	rather weak
• Monopoly on transboundary information flows	rather strong	rather weak
• Basis for rent-seeking activities	rather strong	rather weak
3. International river commission		
• Establishment of the commission	rather difficult	easier
• Resources/ responsibility of the commission	rather weak	rather strong
4. Cross-border activities of private parties		
• Delegation of negotiations to governments	possible	possible
• Transboundary information flows	not possible	possible
• Participation in foreign legal/administrat. acts	not possible	possible
• Direct negotiations with other private parties	not possible	possible

With regard to the analysis outlined above, what does empirical evidence tell us about the actual internalization of external effects in international river basins? Two case studies covering the Euphrates and the Rhine rivers highlight the empirical evidence of some of the theoretical arguments.

4.3 Loosely Coordinated Governments Along the Euphrates, Private Ties Across the Rhine

For centuries both the Euphrates and the Rhine have been the object of major conflicts. Today the Euphrates is still the cause of great international tensions, since Turkish plans to build huge dams on the upstream part of the river threaten to cut off the downstream countries' water supply,[22] causing major environmental

[22] Turkey wants to increase its share of the Euphrates water from 10% in 1991 up to 50% in 2040, with Syria extending its share from 15% to 33%. Iraq, which received

problems in Syria and Iraq.[23] Since 1962 the governments of Turkey, Syria and Iraq have held 16 rounds of discussions without making any progress towards an agreement over a distribution scheme for the river's water.[24] All three governments are playing a complex bargaining game where water distribution is not the only issue. Other issues, such as domestic policy, Kurdish freedom fighters, regional power and the Gulf War have all been on the agenda over the last ten years alone. These endless rounds of intergovernmental small-talk are extremely costly: cooperation could lead to huge efficiency gains of much more than US$ 10 billion annually.[25] The willingness to accept such huge opportunity costs of a non-agreement is quite astonishing, if one considers that actually only 21% of the Euphrates waters is contested.[26] In line with the theoretical analysis outlined above, all governments have used systematically the possibility offered by international law to monopolize negotiations concerning the Euphrates' waters - there is no other forum for such negotiations besides this one ineffective intergovernmental committee. But the governments have not only monopolized negotiations, they are also controlling transboundary (and even domestic) information flows quite effectively: in Iraq, dam construction schemes are an official secret and both Iraq and Syria are considered authoritarian states.[27] Although Turkey is regarded as 'partly democratic', the Kurds, who live around the dam construction sites in southeastern Anatolia, are being suppressed.[28] However, negotiations over the Euphrates have been and are being dominated by the governments involved. They also hold a monopoly on transboundary information flows. Although cooperation could lead to huge efficiency gains, benefiting all parties, the governments refuse to cooper-

75% of the water in 1991, will be left with only 17% in 2040. See Kolars and Mitchell 1991, p.268.

23 The Turkish dam scheme will not only lead to water scarcity and the usual environmental problems of huge dams, e.g. a drop of the ground water level downstream. As the Southeastern Anatolian Project includes a systematic industrialization, the reduced amount of water reaching Iraq will also be likely to be much more polluted than today.

24 For a detailed description of the cooperation process, see Durth 1995.

25 Cooperation would open the doors for Turkish plans to sell water in the Arab world (estimated annual efficiency gains US$ 9 billion), electricity to Syria and a share of the planned Anatolian agricultural production (US$ 6 billion) to Syria and Iraq. As both downstream countries will soon depend on food imports, Turkey has major comparative advantages in this field (as well as in the field of electricity production). See Durth 1995.

26 The river flows with 950 m^3/sec approximately. Iraq and Syria claim 700 m^3/sec - this means they accept Turkish claims for 250 m^3/sec. Turkey guaranteed Syria in a bilateral protocol in 1987 an average flow of 500 m^3/sec., and Syria agreed with Iraq to share the 500 m^3/sec. (58% for Syria and 42% for Iraq). This means that Syria and Turkey accept a 'just' Iraqian share of the Euphrates of 210 m^3/sec., and that Turkey and Iraq accept a 'just' Syrian share of 290 m^3/sec. If one subtracts the mutually accepted 'just' shares of the three parties from the entire stream, only 200 m^3/sec., or 21%, are still contested. See Durth 1995.

27 Hauchler 1993, p.100.

28 Hauchler 1993, p.100. Ironically, Turkey and Iraq, which were struggling over the Euphrates, in a security protocol even allowed each other to pursue its own citizens on the other state's territory ('hot pursuit'). See Durth 1995.

ate. They ignore potential efficiency gains and maximize their discretionary freedom instead. Given the national rivalries in the region, it may be assumed that governments suspect negative effects on their popularity if cooperation leads to compromise. Private groups play no part at all in the dispute.

In contrast to the Euphrates, the cooperation in the Rhine Basin is one of the most successful ones world-wide, and the International Commission for the Protection of the Rhine is being studied as a model for other river basins. However, so far little attention has been given to the role private interest groups have played in solving the river's pollution problems.[29] The potential for private action to solve transboundary externality problems may be illustrated by the Port of Rotterdam's action program to stop contamination of the river's sediments. In the early 1980's the sediments in the Port of Rotterdam were found to be so contaminated with heavy metals and other substances from upstream polluters that special deposition became necessary. The Port of Rotterdam had to spend 200 million fl. for a special deposit site with a storing capacity of approximately 15 years. In order to avoid similar expenses after those 15 years, the Port started a ten year action program involving up to 50 employees and comprising action in three different fields: first, a consulting firm was charged with establishing accounts of contaminants for the whole River Rhine including all important tributaries in the Netherlands, Germany, France and Switzerland. Secondly, a foreign public relations bureau was paid to publicize the port's problem and pursue its claims in the upstream countries. Thirdly, lawyers were engaged to work out comprehensive legal expertise which was then used as a basis, both for direct negotiations with private upstream polluters and for transboundary civil law proceedings. In addition, the port succeeded in promoting its claim as one of four main objectives of intergovernmental cooperation in the Rhine Basin.[30] The Port of Rotterdam's activities show that externalities lead to economic damage which may be strong enough for private organisations to stand up and fight for a clean river. The legal framework inside the European Union provided the basis for most of the port's transboundary action: it could pursue its economic interest not only in the Netherlands but also in the upstream countries using a cleverly planned strategy. The strategy aimed at influencing domestic and foreign public opinion, politicians and polluters with different and appropriate means. When the programme came to its conclusion, the Port of Rotterdam had individual contracts with nearly all the major polluters in the Rhine Basin, promising to reduce their emissions substantially by the year 2002.

Though the Port of Rotterdam had a very strong single preference for a clean Rhine, transboundary actions for other private interests in the Rhine Basin are not unusual either if collective action may not be quite so easily organized among bigger groups. One famous case is the legal suit of Dutch market gardeners against a French potash mine. After 14 years the proceeding ended in 1988 in a victory for the gardeners, and only a few weeks later Dutch officials refused to discuss the matter in the intergovernmental river commission any longer: private transbound-

[29] For a detailed description of the cooperation process in the Rhine Basin, see Durth 1996a. The potential for 'privatization' of international externalities is discussed in Durth (1996f).

[30] See IKSR 1987, p.5.

ary claims had relieved intergovernmental talks. Another famous example for transboundary private action in the Rhine River Basin is provided by the waterworks. As early as 1970 waterworks in all river states formed a powerful international alliance in order to influence public opinion and the objectives of international cooperation in their favour. The waterworks have equal or at least complementary goals - upstream and downstream waterworks need a clean Rhine - and a common bureau in Amsterdam, which is largely financed by Dutch waterworks. By publishing the Rhine Memoranda, which was based on scientific expertise, the waterworks alliance has repeatedly influenced intergovernmental cooperation with great success during critical moments of the river's restoration.[31]

The empirical evidence from the Euphrates and the Rhine largely fits the theoretical analysis of typical features from sections 1 and 2. The following Table 4.2. summarizes the case studies.

Table 4.2. Empirical evidence from the Euphrates and Rhine rivers

Empirical evidence	Euphrates	Rhine
1. Requirements for international cooperation		
• Binding commitments	not existing	existing
• Convergence of ideas of a fair utilization	no convergence	convergence
• Negotiation powers of the riparian states	very unequal	balanced
2. Role of governments		
• Monopoly on transboundary negotiations	very strong	no monopoly
• Monopoly on transboundary information flows	very strong	no monopoly
• Basis for rent-seeking activities	very strong	rather weak
3. International river commission		
• Establishment of a commission	only a technical committee	Rhine River Commission
• Resources/ responsibility of the commission	extremely weak, inactive	strong, very active
4. Cross-border activities of private parties		
• Delegation of negotiations to governments	only channel for private interests	one channel among others
• Transboundary information flows	no private information flows	abundant information flows
• Participation in foreign legal/administration acts	no participation	participation
• Direct negotiations with other private parties	no direct negotiations	intensive direct negotiations

31 Durth 1996a.

4.4 Policy Advice for Solving International Upstream-Downstream Problems

Europe has a rich experience in enduring conflicts over transboundary rivers as well as in inventing cooperation schemes that allow the realization of the efficiency gains from the internalization of externalities.[32] The most important lessons learnt in the integration process of Europe after 1945 may be summed up in ten rules for solving upstream-downstream problems:

1. Internalization of transboundary externalities leads to efficiency gains. Economically, the realization of these efficiency gains is the major reason for cooperation between the riparian states. Publicizing the expected efficiency gains can increase the willingness to cooperate.
2. These efficiency gains can only be realized if all riparian states are willing to make binding commitments. Credible sanction mechanisms, such as the 'exchange of hostages' or interdependencies, help to make binding commitments possible.
3. The efficiency gains will only be realized if different ideas on a *fair* utilization of the common resource converge. In cases of huge differences, it may be advantageous to reach a consensus on *how* a fair outcome can be achieved instead of discussing possible fair outcomes *themselves*.
4. Negotiations may also be broken off if there are big differences in negotiation power. A powerful party determined to realize the efficiency gains may deliberately weaken its own position in order to ensure smooth negotiations.
5. Any international cooperation needs the consent of all governments involved. Especially in non-integrated areas where there is no alternative to intergovernmental negotiations, governments will only allow for internalisation if they regard cooperation as advantageous to themselves. Anyone looking for a solution of upstream-downstream problems should, hence, analyze the costs and benefits for all the governments concerned. Costs and benefits are not limited to purely economic factors but include political effects as well. Empirically, the anticipation of effects on the popularity and the prestige of governments plays a central part in promoting or hindering international cooperation.
6. Internalization activities should not be limited to governments. On the contrary, all parties having an interest in the internalization process should be allowed to participate in transboundary information flows and start their own negotiations with foreign parties.
7. An international river commission may be helpful to the effect that it reduces transaction costs of and weakens government monopolies on transboundary information flows and international negotiations. When established, a commission should explicitly analyze the ecological and economic consequences of internalization, since this information will help to maximize efficiency gains to increase the governments' willingness to cooperate.

---

[32] Durth 1996b.

8. Economic interests of private parties can take on an important role in the internalization process. Activities from private parties suffering from external effects should be tolerated in their own country as well as in foreign countries. In order to encourage such activities private interests can be strengthened by establishing property rights on national and, ideally, supranational level. Interfaces between national laws could facilitate claims by foreign private parties. Furthermore, the organisational capacity of private parties suffering from external effects may be increased by direct public support focusing on typical problems of collective action.[33]
9. Cooperation must be viewed as a long-term process, the terms of which are designed to reach a specified aim (internalization). External shocks contrast with this somewhat pedestrian rate of progress, as they can cause a radical and unforeseen change in the cost-benefit calculations of individual participants. Such shocks are of particular importance because they can help to overcome political and economic inertia standing in the way of internalization; often they are *windows of opportunity*.
10. In Western Europe the economic and demographic changes began much earlier than in most other regions of the world. As a result, experience has been accumulated over a longer period and sophisticated principles of cooperation have been developed. From the viewpoint of non-integrated regions the European Union can be interpreted as a vast storehouse of experience. The impact of this institutional experience on Europe became evident only a few years ago: although in 1990 the German and Czech authorities had practically no experience of cooperation on the Elbe River, they were able to assimilate and apply the wealth of experience of the Rhine Commission relatively easily; within four years, rather than forty, the Elbe River states had caught up with their counterparts on the Rhine.[34]

References

Barret, S. 1992: Self-Enforcing International Environmental Agreements. CSERGE Working Paper GEC 92-34, London.

Biswas, A. 1993: Management of International Waters: Problems and Perspective, in: Water Resources Development, Vol. 9, No. 2, pp.167-188.

Caponera, D. 1987: Patterns of Cooperation in International Water Law: Principles and Institutions, in: Utton, A.; Teclaff, A. (eds.). Transboundary Resources Law. Boulder, London. pp. 1-27.

Coase, R. 1969: The Problem of Social Cost, in: The Journal of Law and Economics. Vol.12, No.1, pp.1-44.

Durth, R. 1995: Internationaler Streit ums Wasser: Konflikt- und Kooperationspotential am Euphrat, in: Orient 36/4, pp.649-662.

Durth, R. 1996a: Der Rhein - ein langer Weg zum grenzüberschreitenden Umweltschutz, in: Aus Politik und Zeitgeschichte B7/96, pp.38-47.

[33] Olson 1991.

[34] Durth 1996c.

Durth, R. 1996b: Grenzüberschreitende Umweltprobleme und regionale Integration. Zur Politischen Ökonomie von Oberlauf-Unterlauf-Problemen an grenzüberschreitenden Flüssen. Baden-Baden.

Durth, R. 1996c: Grenzüberschreitende Umweltprobleme an der Elbe - Gestern Eiserner Vorhang, heute ein europäischer Fluß, in: Deutsche Studien 33/2, pp.123-139.

Durth, R. 1996d: Zusammenarbeit bei internationalen Oberlauf-Unterlauf-Problemen und Integration, in: Zeitschrift für Umweltpolitik und Umweltrecht Nr. 2; pp.183-208.

Durth, R. 1996e: European Experience in the Solution of Cross-Border Environmental Problems, in: Intereconomics 31/2, pp.62-67.

Durth, R. 1996f: Wie lassen sich grenzüberschreitende Externalitätenprobleme lösen?, in: Aussenwirtschaft, 51/IV, pp.539-557.

Hauchler, I. (ed.) 1993:. Globale Trends 93/94, Frankfurt.

IKSR 1987: Aktionsprogramm 'Rhein', Straßburg.

Johnson, S.P.;Corcelle, G. 1992: The Environmental Policy of the European Communities. London, Dordrecht, Boston.

Kiss, A.; Shelton, D. 1991: International Environmental Law. New York.

Kolars, J.; Mitchell, W. 1991: The Euphrates River and the Southeast Anatolia Development Project. Edwardsville.

Kuhl, H. 1987: Umweltressourcen als Gegenstand internationaler Verhandlungen. Frankfurt.

Kunig, P. 1992: Nachbarrechtliche Staatenverpflichtungen bei Gefährdungen und Schädigungen der Umwelt, in: Dolzer, R. et al. 1992: Umweltschutz im Völkerrecht und Kollisionsrecht. Berichte der Deutschen Gesellschaft für Völkerrecht Heft 32. Heidelberg. pp.9-56.

Lammers, J.G. 1984: Pollution of International Watercourses. Leiden.

Luge, I. 1989: Haftung als notwendiger Teil des internationalen Meeresumweltschutzes: Der Entwurf der HNS Convention und sein Umfeld. München.

Mäler, K.-G. 1991: Incentives in International Evironmental Poblems, in: Siebert (ed.) 1991: Environmental Scarcity: The International Dimension. Tübingen. pp.75-93.

Meaows, D. et al. 1992: Die neuen Grenzen des Wachstums. Stuttgart.

O'Connell, M. 1992: Enforcing the New International Law of the Environment, in: German Yearbook of International Law 35, pp.293-332.

Olson, M. 1991: Die Logik kollektiven Handelns. Tübingen.

Paleologou, E. 1993: La protection des eaux par le droit communautaire; Cours d'été en droit et politique communautaires de l'environnement. Louvain-La-Neuve.

Rest, A. 1986: Luftverschmutzung und Haftung in Europa: Anspruchsmöglichkeiten auf nationaler, internationaler und völkerrechtlicher Ebene. Kehl.

Sandler, T. 1992: Collective Action. Michigan.

Ströbele, Wolfgang J. 1991: Institutional Arrangements for Transfrontier River Pollution, in: Siebert, Horst (ed.) 1991: Environmental Scarcity - The International Dimension, Tübingen. pp.139-159.

5 Contested Waters: Dividing or Sharing?

Ulrich Küffner

In desert regions and arid lands the lack of water restricts human activities, plant growth and animal life, but it also shows that life can exist with very little water. In the past, there appeared to be a balance between water and human activities, city life and agriculture in the dry areas. But population growth, industrial development and irrigation, especially a rapid expansion of irrigation through modern techniques, have overexploited the water resources in many arid countries. Rivers are running dry and ground water tables are falling where water is abstracted without regard to natural replenishment or recharge of the water resources. And the water demand is still rising.

International conflicts over water are well known. Conflicts over water have occurred between many countries in all climatic regions, but between countries in arid regions they appear to be unavoidable. Claims over water have led to serious tensions, to threats and counter threats, to hostilities, border clashes and invasions. The Middle East is the prime example for a region where all these problems can be found.

Where a precious resource is claimed by several or many contesters, the ideal solution seems to be its just and equitable distribution, clearly defining the shares for the various parties. Interest groups, lawyers, legislators and academics have tried to find answers to the question of equitable distribution of water, but the goal remains elusive. The proposals are intellectually stimulating but hardly useful in politically charged negotiations where power and the influence of powerful allies dominate.

The Middle Eastern countries have made some progress towards the distribution of water, but an overall solution through dividing water resources among all claimants does not appear to be reachable. Further disputes and conflicts are to be expected if the governments proceed within the traditional framework of trying to define just and equitable shares of water.

This paper explores the possibility of another solution to the problem: the sharing of water among the various parties instead of dividing it into defined portions or allocating certain quantities to communities or states.

5.1 Riparian Conflicts

Water has been the cause of many conflicts. Farmers are known to have fought, even killed each other over access to water. It is interesting to note that the expression 'rivals' stems from the Latin word 'rivales' for people living, and confronting each other, along a river. Such fights still occur in farming communities. Water rights are also disputed among municipalities that wish to secure sufficient water for their customers. Provinces or states within countries have fought and are fighting in courts for the right to larger shares of water from rivers and ground water aquifers. However, international conflicts have received much more attention and are undoubtedly much more difficult to solve.

The seriousness of the conflicts has even led some commentators to predict water wars. The same authors usually quote politicians who threatened neighbouring countries with war if they initiated some water development works which would reduce the flow of water. Should such threats be taken seriously? To find an answer to this question and explain the problems of water allocation, the situation in some of the most affected river basins in arid regions will be reviewed briefly.

Most large rivers cross national borders, and great lakes and ground water aquifers are shared by several countries. In arid regions the main river basins are truly international: the Nile and Niger in Africa, the Indus and the Euphrates-Tigris in Asia and the Rio Colorado and the Rio Grande in North America. The Nile is shared by ten countries. Egypt and Sudan claim currently the whole flow of the river under a bilateral agreement. The upstream riparians, particularly Ethiopia, rejected these claims and refused to enter into negotiations with Egypt and Sudan under those conditions.[1] Many attempts have been made to bring all the riparians together, but only some agreements on joint planning efforts have been reached. Lacking the possibility of major water development projects, especially in Ethiopia, which would claim a substantial part of the Nile flow, the upstream governments adopted a rather passive position, reserving their rights to claim larger shares of the Nile flow, but not really challenging Egypt, the major user of water.

In the Niger Basin with its eight riparians, Nigeria is by far the most powerful country and largest user of water. It has dominated all past efforts to develop the river, but has conceded traditional water uses in the upstream countries and minor additional abstractions without any formal arrangements. Here also, the economic weaknesses of the upstream riparians have prevented them from attempting any major developments and thus challenging Nigeria's dominating role.

The Indus River is shared mainly by India and Pakistan. After the separation of India and Pakistan, the use of the Indus water soon became one of the major, if not the most important disputes between the newly independent countries. The water agreement which was eventually reached between the two governments is often mentioned as the prime example of how an agreement can withstand many conflicts and even several wars.

---

[1] See Tilahun 1979.

In the Euphrates-Tigris Basin, Turkey is the predominant power which has abstracted an increasing share of water in recent years for a large scale irrigation and hydropower development in southeastern Turkey.[2] Although Iraq and Syria protested the unilateral Turkish actions, they were unable to challenge Turkey directly - unless the support for Kurdish uprisings in Turkey is seen as a reaction to the Turkish intervention. It also has to be recognized that the abstractions of the Euphrates' water by Turkey have not produced any serious water shortages in Iraq and Syria which traditionally have used large volumes of water for irrigation. Basin-wide cooperative efforts have unfortunately been "foundered on the rocks of conflicting interests among upstream and downstream nations".[3]

In North America, the Rio Colorado and the Rio Grande are both shared by Mexico and the United States of America. Initially, agreements were reached between the two countries that specified quantitative abstractions. However, the increasing pollution of the Rio Colorado made the water for Mexico, the downstream riparian, practically useless. After many protests and lengthy negotiations, the U.S. government agreed to build a major water treatment plant near the border with Mexico to deliver not only the agreed quantity of water but also water of an acceptable quality to its neighbour.

One of the most contested river basins, if not the most contested one, is the Jordan River Basin where five peoples claim parts of the water: Israelis, Jordanians, Lebanese, Palestinians and Syrians. Only two of the five, the Israelis and Jordanians have reached an agreement on how to partition the water. No agreement is in sight with the Lebanese and the Syrians, and Israelis and Palestinians have signed only a tentative agreement on the use of ground water underlying the West Bank, i.e. the Oslo Agreement (1995). While the Egyptian-Sudanese agreement on the use of the Nile water includes a provision on how to deal with claims from other parties, the Israeli-Jordanian agreement does not include such a provision. The two governments divided the flow of the Jordan River among themselves, although there are claims by the Lebanese, Palestinians and Syrians. How such claims could be accommodated in the framework of the bilateral Israeli-Jordanian agreement remains to be seen. One suspects that the initial reaction to any claim would be negative and that there would be suggestions to seek a portion of the share of water from the other party. The parties are thus far from a multilateral agreement.

In spite of the rather negative outlook for a lasting solution in the Jordan Basin, several attempts have been made in recent years to assist the parties in their search for a settlement. The latest attempt is a comprehensive study supported by the German government exploring future water needs and the possibilities for mobilizing additional water. The study has shown that the current water use exceeds the available resources in Jordan, and that Israelis and Palestinians will soon need additional water if their projections of future water needs turn out to be correct. The needs of the agricultural sector will be decisive since irrigation uses by far the largest volume of water. Will this allocation continue to a sector which is rapidly becoming less and less important for the economies and for employment? The

2 See in this volume Scheumann.

3 See Kolars 1994.

governments involved will have to face the powerful agricultural lobbies and the emotional attachment to agricultural development in the Middle Eastern societies.

5.2 Traditional Solutions

Many international conferences and international organizations have urged governments to cooperate in managing the water in international river basins.[4] But how can water be allocated in an international river basin among the riparians? Many attempts have been made to find answers to this question, to determine a just and equitable distribution, and to establish guidelines for negotiations, but the goal remains elusive. Interest groups, lawyers, legislators and scholars have made numerous proposals, and courts have ruled on this subject, but a solution satisfactory to most parties has not been found. The Helsinki Rules of 1966, accepted by many governments, set out principles for the distribution of water, but they are overly general and thus of little practical use. The same can be said of the many proposals that have been made to solve the problem. They are interesting and intellectually stimulating, but they have not helped any party locked in protracted negotiations.

Many countries have signed international agreements over the use of water. As mentioned above, Egypt and Sudan agreed on how to divide the flow of the Nile between the two riparians. In their water agreement of 1959, they divided the total average annual flow among themselves, but introduced a provision that they would address any claim from upstream countries jointly, and if they agreed to concede any water to other claimants, they would give up equal parts from their current allocations. Any flow increase due to upstream development work would also be split in equal parts.[5]

India and Pakistan found a formula for dividing the water in the Indus Basin. It took many years of negotiations and the intervention of influential outsiders to reach a solution. While the agreement has survived many conflicts, even open wars between the two countries, the solution itself, i.e. the partition of the flow regime through massive investments in canals and dams, has been criticized as a wasteful use of resources, since a collaborative agreement of sharing the river would have been much more economical.

Israel and Jordan signed a treaty recently,[6] which determines how the two countries intend to distribute the flows of the Jordan and the Yarmuk rivers, i.e. the most important tributary of the River Jordan.

The agreements show little regard for the principles laid down in the Helsinki Rules or the proposals made by others on the issue of just and equitable allocation. Contrary to the generally accepted principles, most agreements do not apportion water according to needs and development opportunities, but sanction the status quo of water allocation with some modifications to give the weaker parties an

4 United Nations 1977; World Bank 1993.
5 See Waterbury 1979.
6 Treaty 1994.

incentive to sign the treaty. Political, economic and military power are clearly more decisive than general allocation principles.

The international agreements that have been signed are usually bilateral even where many parties claim water from the same river. This also is contrary to the recommendations from experts and international commissions that recommend comprehensive basin agreements. The treaty between Egypt and Sudan is a classic example. The two countries divided the flow of the Nile River among themselves, as mentioned above, without consulting the upstream riparians or allocating parts of the flow to them.

The basic allocation principle of all the agreements is a volumetric division of an average flow. The calculation of the average flow depends of course on the time series considered, and actual flows can differ substantially from the agreed average. Most treaties do not even consider such differences thus inviting serious disputes in case of flows much below the agreed average. Such events, as well as changes in the conditions assumed during negotiations, population growth, economic developments and the like, will necessarily lead to conflicts and calls for renegotiations from the parties that feel at a disadvantage.

In some cases, the agreements draw a distinction between dry and wet season flows. Average low and high season flows are the basis for the allocation between the parties as in the case of the recent Israeli-Jordanian agreement. These provisions reduce potential disputes since dry season flows represent the critical conditions in arid countries.

The most appropriate quantitative distribution principle would be a specification according to percentages of actual yearly or monthly flows, which would make certain that the parties would receive equitable shares of the available flows. However, traditional allocation criteria still predominate. Some agreements even call for a fixed allocation to one party and leave the remaining flow to the other(s). The Israeli-Jordanian agreement is an example of this type of allocation which guarantees a certain volume of Yarmuk River water to Israel and leaves the remaining flow to Jordan. This specification favours Israel in dry years and Jordan in wet years. It also includes an element of risk for Jordan as the upstream riparian, Syria, may abstract more water in the future and thus reduce the flow available to Jordan. And Jordan needs to build storage reservoirs to capture the very irregular wet season flows. Unfortunately, reservoirs that can only be filled in wet years are much less economical than those which receive water regularly.

Finally, some agreements consider the question of water quality. The U.S.-Mexican agreement on the use of the Rio Colorado mentioned above added such provisions at a later stage when the increasing pollution through water use in the USA made the water practically unusable south of the U.S. border. Such provisions should be considered in all agreements, but very few have included them, leaving the downstream riparians at the mercy of the upstream countries.

The partition or allocation of water through distributing it volumetrically is the usual method in agreements within countries. To solve conflicts arising from disagreements over the allocation of water 'conflict resolution techniques' have been developed. In the United States in particular, conflict resolution techniques have become popular to reduce the cost of lengthy litigation processes. These techniques aim at bringing the interested parties, the stake holders in the water sector, together at an early stage to familiarize them with the main issues and the concerns of all the parties involved. The active participation of the stake holders in finding solutions, and the reconciliation of the various claims with the help of facilitators is nowadays a well accepted procedure which is credited with the successful resolution of many conflicts. In the international sphere such techniques have until now been less successful, although many have tried to play the role of facilitators. In the Nile Basin, for example, the United Nations Development Programme (UNDP) has devoted many years of efforts to bring the riparians together but with little success.

While most international agreements are bilateral, multilateral or comprehensive basin agreements are advocated by independent water experts and international organizations, as mentioned above. One can speculate why such multilateral agreements have not been reached. Skeptics may argue that the most powerful riparians wish to use their position in one-on-one negotiations, preventing the weaker parties from presenting a united front against the stronger country. This consideration may play a role, but the complications of a multilateral approach can not be underestimated. Even joint basin planning efforts, which have been carried out in several international river basins, have shown the difficulties of reaching a consensus on a common approach. Examples are the planning activities in the Kagera, Lake Chad, Niger and Mekong river basins. Often the expectation of favourable funding of joint projects brought governments together, but when such expectations did not materialize, the various governments did not feel bound to the agreed plans and pursued their own interests without regard to the interests of the other countries.

In spite of their deficiencies, existing international agreements have resolved some serious disputes over water. However, some of the agreements are in need of renegotiation or modification because they are based on a rather rigid allocation formula and do not provide for adjustments to changing conditions. Are there better solutions? This is a question that should be explored since conventional agreements are still seen by many as the only way to solve international water conflicts. The following sections will analyze the potential for alternative solutions.

5.3 Towards an Alternative: Sharing Instead of Dividing

In the late 19th century, municipalities and large companies in the Ruhr Valley, i.e. the main German industrial district, confronted the problem of insufficient water for the population and the rapidly growing industry. They also confronted each other as they were trying to tap water from the Ruhr River and the tributaries of the Ruhr south of the main valley. Instead of developing water supplies individually - water supply and sewage services being a municipal obligation - they decided to pool their resources and form an association, i.e. the Ruhrtalsperrenverein. The association was assigned the task of developing water supplies for the municipalities of the Ruhr Valley, building reservoirs, water treatment plants and supply lines to the demand centers. The activities of the association were complemented several years later by the work of the Ruhrverband responsible for sewage collection and treatment.[7] The Ruhr district thus shares the water resources of the region, providing equal access to water for all consumers. This model was not followed in many areas of Germany, since the country enjoys plentiful water resources and few regions are as densely populated as the Ruhr Valley. Only larger cities like Hamburg, Frankfurt and Stuttgart needed water supplies from distant sources for which water carriers were built, usually operated by separate companies.[8] However, in the former East Germany where the political and economic role of the municipalities was substantially curtailed, regional water associations were established to take advantage of the economy of scale of water works. The French system of river basin committees and agencies, which was established for the whole country after the second world war, also drew on the experience of the Ruhr district.

In the United States, when several local water agencies in the Washington metropolitan area faced the need to develop additional water supplies for the various communities, they realized that each one needed very costly new water investments. Since they found out that separate water development programmes would have been many times more expensive than a joint development scheme for all communities, they agreed on joint water development works that now deliver water to the various municipal and county water agencies.

Complete mergers of separate water agencies have taken place where large cities were formed, as in Buenos Aires and Mexico City, with over 10 and 20 million inhabitants. Smaller municipalities with their water agencies were merged with the core cities resulting in unified systems that provide water to all citizens in an equal manner and at the same tariffs.

A few countries, mostly city states and Israel, built country-wide water systems that use all water resources to serve their citizens through a common network of canals and pipelines. Here again, all consumers have equal access to water and pay the same tariffs.

There are some lessons that can be drawn from these examples. First of all, cooperative arrangements and mergers of water agencies have eliminated conflicts over water between water agencies, municipalities or regions. Previously, water agencies or water companies and municipalities faced each other in court or in bidding wars over water, which often delayed the development of water resources. After the completion of the cooperative arrangement or merger, water resources became a joint resource, and water projects could be implemented without any challenges from competing parties.

Secondly, the cost of water development and management has been considerably reduced. This is a very powerful argument for consumers since it will lower the cost of water, or at least prevent higher costs, when separate companies need to implement separate development projects. The partners of an association and merged companies can pool their resources and thus implement joint projects such

---

[7] See Radzio 1988.

[8] See Lehn et al. 1996.

as the construction of dams or well fields and pipelines and water treatment plants, which individual water companies might only build at higher costs or not at all, eventually curtailing water deliveries to consumers. The cooperating agencies or companies need, of course, physical connections to their supply systems to allow the pooling and sharing of resources. In addition to the economy of scale of the technical facilities, cooperative or merged managements allow the streamlining of administrations, maintenance and repair services, billing procedures and the like, which in turn reduce the cost of providing water.

Thirdly, cooperative arrangements and mergers provide water to all consumers on an equal basis. While the services of individual water companies differ in terms of continuity of water delivery, water quality, pressure and price, cooperative arrangements and mergers result in equal and usually much improved services for all. However, the allocation of water to different sectors remains contentious within unified systems, particularly the allocation of water to agriculture since agricultural users are usually successful in obtaining water in larger quantities and at lower prices than domestic and industrial users.

Furthermore, cooperative arrangements can take many forms. They might range from an overall coordination of water activities and the joint management of resources with bulk supplies to individual water companies, to the complete merger of water companies. Government control over the resources remains essential. Water companies, state owned or private, are natural monopolies due to their control of water distribution networks, and have the tendency to exploit their particular advantages and should therefore be under the control of a strong supervisory agency.[9]

Cooperative arrangements in the water sector are quite limited in contrast to other sectors. Energy companies have established wide ranging networks across borders of many countries with sharing arrangements for power generation and transmission. Telecommunications companies are actively pursuing the creation of world wide systems. And large companies in many sectors, such as automobile manufacturing, chemicals, oil or pharmaceuticals are becoming more and more international in their operations. Water companies seem to shy away from such arrangements. Only a few large water companies , especially French and English companies are investing in other countries where water companies are being privatized and open to foreign investments. These cooperative arrangements in the water sector are encouraging but they are quite modest and do not really span borders in the way of sharing water resources.

A futuristic vision for the Middle East was developed by Kolars, foreseeing „a time when hydrologists, agronomists, power engineers and diplomats would meet at a centre which would incorporate among its many facilities satellite receiving stations transmitting not only immediate satellite images of weather conditions but also the readings of remote stations (...) these, in turn, would be synthesized with archival records by GIS (Geographic Information System) technology into a comprehensive picture of water availability, probabilities, and water needs throughout the international region. Diplomatic brokers might then bargain and exchange

[9] See The World Bank 1994.

supplies of electric power, stored water, petroleum, or available food and industrial crops".[10] The vision, though futuristic regarding the use of modern technologies, is not far from reality considering the buying and selling of water between national entities. But water sharing would need a different approach.

5.4 Regional Water Sharing

Is the concept of water sharing across national borders practical and could it be applied to arid countries?

Water is already being channeled across national borders. City states like Singapore and Hong Kong buy water from neighbouring territories. Islands in the Caribbean region depend on the import of water by tankers or barges to satisfy the water needs of the growing numbers of tourists. Water was occasionally transported from Scotland to Mallorca when droughts in Southern Spain prevented supplies from the Spanish mainland. A comprehensive agreement was signed between South Africa and Lesotho for the Orange River, guaranteeing a regulated river flow to South Africa against the payment of royalties and the financing of water infrastructure by South Africa in Lesotho. The basic concept is thus not completely new. Water is traded, sold from places where it is available to locations with urgent demand. The cost can be high, but where water is needed people are willing to pay the price even for 'imported' water.

But the trading of water across national borders is only the supply of some limited quantities against payment. It shows that transnational arrangements are being practiced, with economic advantages to buyers and sellers. A first step of cooperative arrangements of neighbouring countries could, therefore, be an interconnection of water systems, with the aim of buying and selling water when the need arises, or the surplus in one system allows the transfer of some water to an adjacent system. This solution would only require technical connections through canals or pipelines, and contractual agreements regarding the water trade, specifying the price of water, its quality, could possibly be differentiated according to dry and wet seasons.

An association of water agencies or companies from neighbouring countries could begin with agreements on the sale and purchase of water. Full cooperation would, of course, require much more. The advantages appear to be obvious: the elimination of conflicts, a reduction of costs, and equal access to water for all consumers as discussed above, in connection with the possible cooperation or merger of national water companies. But undoubtedly, many problems would have to be solved. Even the merger of national companies usually requires lengthy negotiations and new contractual and, possibly, legal provisions.

The main problems of a complete cooperation or merger of water companies of different countries will be the sharing of the resources; the establishment of an interconnected water system; changes of legal provisions; the creation of a com-

[10] Kolars 1994, pp.90-91.

mon organizational set-up to manage the system; an agreement on tariffs in different currencies and the question of water allocation to the main sectors in the different countries.

The sharing of the water resources would be a fundamental change of the current concept of water rights, which often belong to the state and would have to be transferred to a new entity. In many countries, especially arid countries, there is an emotional attachment to water because of its preciousness, its scarcity and its symbolism which is manifest in religious ceremonies. The transfer of water rights, perceived to be essential to the well being of a society, will be difficult. Changes in the legal provisions regarding water rights would undoubtedly be needed in most countries. While advocates for maintaining national water rights would surely oppose such a solution, consumers who would be supplied by an interconnected network with resources from different areas, would not notice any difference, just as consumers who depend on interconnected electrical systems with power generating plants in several countries. The question of sharing water resources would be a political decision which would go against tradition and perceived national interests, but it would bring economic and operational advantages.

Interconnections between systems of neighbouring countries would raise questions of technical feasibilities and costs, similar to the connections of water systems of several cities which agree on a merger of their facilities. It would require connections that would allow sharing resources and transferring sufficient quantities of water in case of shortages or emergencies. Interconnections are thus simply a matter of additional investments which would be part of the cost of the new arrangement.

The changes or adjustments of the legal systems would be more difficult to implement because of the inherent resistance to any legal changes, as well as the many legal provisions concerning water that have been developed in most societies. However, it would probably not be much more difficult than the privatization of water companies, an issue widely debated in many countries.[11] If the establishment of private water companies were allowed, the creation of a bilateral or international water company serving two or more countries would also be possible. But the resistance even to national private water companies should not be underestimated.

The creation of a joint organizational set-up may appear to be difficult, but most countries have some experience in the formation of independent agencies or companies which are charged with the management of water related activities, such as the management of pipelines, reservoirs or treatment plants. Similarly, a joint company or agency could be established as an independent entity with the responsibility for water management. One peculiar difficulty is the need for considering the many variables among which the decision makers would have to choose. They range from an intergovernmental agency - since government agencies dominate the sector in many countries - to a fully private water company. A mixed company, partly owned by the governments concerned and partly owned by the private sector, may be favoured by those who wish to maintain some direct government influ-

[11] See Rivera 1996.

ence. The joint agency or company would need to be independent regarding financial management, the raising of capital and setting of tariffs, and personnel management, and would have to follow clearly defined rules and regulations. In addition, a joint or multilateral regulating and control authority would be essential to guarantee that the tasks be carried out properly and prevent the abuse of monopolistic powers.

The setting of adequate tariffs should be the responsibility of the joint agency or company within a framework set by the regulating authority. Tariffs, however, have become a political issue in many countries because they are often used as a sociopolitical tool. Some politicians argue that poor consumers and certain groups or industries would need support through low (subsidized) tariffs. The support of certain income groups as a tool of socioeconomic policy is not questioned here, but the form of support through generally low water prices - particularly in arid regions - appears to be inherently wrong as it provides the wrong signal to consumers to use more water rather than conserve it, which should be a prime objective. Support to low income groups and certain industries could easily be provided through other means: direct payments or tax rebates - but much better through cross subsidies within the water sector - however, the sector, if financially independent, needs to have cost covering tariffs. Thus realistic tariffs that cover the full cost of water would have to be agreed upon. Realistic tariffs would also reflect location specific conditions, i.e. the cost of conveying water to distant locations and to higher elevations. This should also avoid or reduce the establishment of water intensive industries in such locations. The question of different currencies should not be an impediment since the real cost of water would have to be charged to the consumers as in the case of electricity tariffs, where consumers depend on or enjoy access to power plants in different countries.

The allocation of water between groups of water users, in particular the allocation between agriculture and domestic/industrial water supply is often highly skewed in favour of agriculture, though the share of agriculture will decline rapidly with population growth and urbanization.[12] A coordinated arrangement between countries would have to take this distortion into account. There are various ways to deal with this issue. Agricultural water may be separated from domestic/industrial supplies, and sharing arrangements could start with water for the domestic/industrial sector. This would not be very satisfactory, but would facilitate the start of cooperation. Sharing arrangements could be extended to the agricultural sector when the pricing policies for irrigation water could be unified in the participating countries. The long term goal, rarely reached in any country, would be the payment of the full cost of water by all water users.

---

[12] See The World Bank 1994.

5.5 Sharing of Water: A Solution for the Jordan Basin?

Considering the possibility of the sharing of water in the Jordan Basin, it must be recognized that there are formidable obstacles to closer cooperation. Israelis and Palestinians are still far from agreeing on a number of essential issues. Apart from water, the control of land, the status of Jerusalem, the free movement of Palestinians are issues which need to be resolved. Questions about the reliability of the other party are raised frequently. Therefore, lack of trust is a major issue for any agreement between the two peoples who are struggling to find solutions for their coexistence.

Even the cooperation between Israel and Jordan, codified through a treaty, has not yet developed much beyond the declaration of peace, similar to the relationship between Egypt and Israel. The slogan of national interests is being used to maintain or increase the hold on any resources. But what is really the interest of the average citizen as a consumer? To the water user it matters very little or not at all if water is 'owned' and managed by the state, a municipality or an anonymous association, as long as water of adequate quality is delivered in adequate quantities and without interruptions. Here again, the supply of electricity may be quoted. It will be satisfactory to the consumers as long as it is reliable, no matter where power plants are located and who the owners of generating facilities and distribution networks are. Bilateral or multilateral agreements would facilitate a satisfactory water service through a comprehensive system, capturing all available resources and mobilizing additional water when needed. Today, however, national water agencies focus on the national interests in their plans and projects for improving water supply, showing positions similar to those of the political leaders. This is only natural and it shows that questions about the most attractive technical solutions and the most economic water development still take second place after political considerations.

Currently, public agencies dominate the Israeli, Jordanian and Palestinian water sectors. Israel established a water commission with overall responsibility for the sector and a state owned company in charge of treating and distributing water. The Jordanians created a Ministry of Water and Irrigation under which two agencies deal with water supply and irrigation. The Palestinians formed a water authority and are planning to build their own water distribution network. Such government agencies are not interested in changing their approach and their own role in water management. Most likely, they will resist changes from national to multinational management. They are also not in favour of privatization since it would weaken their position, their power and reduce employment in the public sector. Any change would therefore have to consider how positive elements of the state agencies could be preserved.

Besides the lack of trust and the resistance of state agencies, the importance of the agricultural sector and the tariff question appear to be major issues which would have to be addressed in order to find a solution for a shared arrangement in the Jordan Basin. The agricultural sector, which uses by far the largest portion of water, is seen by many in the countries as a vital element of the economies in the region, although the share of agriculture in the gross national product is continuously decreasing and farmers have difficulties finding workers, forcing them to recruit foreigners. Nevertheless, agriculture is very important, with powerful lobbies and political constituencies. In Israel the high cost of water, falling prices of some products such as citrus fruits and the difficulty to recruit agricultural labour have led to questions about the future role of agriculture.[13] The need to use more and more treated waste water for irrigation instead of fresh water is an additional element that will impose changes on the sector. Previous plans to expand the irrigated area will therefore hardly be implemented.

In Jordan, an expansion of the irrigated area is still part of the government plan, although the scarcity of water and the current overexploitation of the Jordanian water resources should halt any expansion. Marketing and employment problems add to the difficulties of the agricultural sector. Nevertheless, the government is providing subsidies in the form of very low water prices that do not even cover the cost of operation and maintenance of the public irrigation systems. In the Palestinian territories, the irrigated area is very small in comparison to Israel and Jordan. The Palestinian water authority would like to expand the area, but the rather small quantity of water available to the Palestinians does not allow any expansion. Considering the problems which Israelis and Jordanians are facing in the agricultural sector, the small Palestinian irrigated area in its current conditions may be a blessing in disguise, as it does not require any subsidies in contrast to the Israeli and Jordanian sectors, and it will make it much easier to introduce changes which will be needed in the near future. A reduction of the irrigation water subsidies in Israel and Jordan will force farmers to make drastic changes in their operations, abandon water intensive crops and focus on less water intensive and higher value crops. Agriculture would thus reduce its water use in the long term when farmers would have to pay the full cost of water. The Palestinians will not have to go through such a painful adjustment process since the irrigation water is not subsidized at all and farmers are still able to maintain their production.

The determination of water tariffs is politically charged in both the agricultural and domestic/industrial sectors. A merger of the Israeli, Jordanian and Palestinian water agencies and the introduction of realistic tariffs as defined above, would result in considerably higher water charges in certain regions. This would apply to consumers in the Jordanian highlands as soon as they require very expensive additional water pumped from lower elevations to the higher demand centres. Consequently, these Jordanian consumers would have to pay not only higher tariffs but also more for water in relation to their (lower) per capita income because of the unfavourable location. Today, higher water tariffs are considered unacceptable by many Jordanian politicians who demand that water tariffs be kept low because of social and health reasons. This raises the question of who should pay the necessary subsidies to keep tariffs low. As argued above, water charges should cover the real cost of water, especially in arid countries where water conservation should be encouraged and wastage prevented. While the totality of water charges should pay

[13] See Lonergan and Brooks 1993.

for the costs of the sector, support to weaker income groups could be provided through cross subsidies within the water sector.

Although the obstacles to real cooperation in the Jordan Basin appear difficult to overcome, there are some indications of a more cooperative spirit among water experts. A group of Israeli and Palestinian engineers and economists interested in water have been meeting during the past years, with the support of some foreign experts, to develop ideas for a joint management of the aquifers shared by the two peoples.[14] Their work, published in a series of conference papers, shows the opportunities and advantages for real cooperation in using, preserving and improving a resource which is undoubtedly common, no matter how the two parties wish to divide it.

An economic water model was developed by American, Israeli, Jordanian and Palestinian scholars. The Harvard Middle East Water Project[15] showed that the quantity of water under dispute in the Jordan Basin is relatively small and its value is low in comparison with the economic potential of the competing parties. The model also shows that existing water resources could satisfy the domestic/industrial water demand until the year 2020. Agricultural demand, however, is seen as excessive, a demand which is increased through subsidies which lower the cost of water for irrigation.

Another encouraging effort is the above mentioned German government supported study of the long term water needs in the Jordan Basin.[16] Teams of Israeli, Jordanian and Palestinian experts are working according to the same terms of reference with each other to determine joint solutions to satisfy the future water demand of the three peoples. While this study did not necessarily lead to joint investments or long term cooperative activities, it brought experts from the three parties together and demonstrated the potential for multilateral activities and the economic attractiveness of joint solutions.

The important task of improved water data collection and processing is the objective of the Middle East Water Data Bank Project, which is supported by the European Union, the UNDP and several western government agencies. Both the Middle East Water Data Bank Project and the German government supported study are carried out within the framework of the multilateral working group on water of the Middle East Peace Process. The working group has sponsored a series of studies on water conservation and on 'new and additional water' for the region. While no breakthrough has been achieved, the fact that Israelis, Jordanians and Palestinians are collaborating with each other, with the support of the multilateral group, is an important step forward in comparison with the situation a few years ago.

A cooperative arrangement for sharing water in the Jordan Basin does not appear to be a realistic solution under the present conditions, no matter how advantageous it may be economically and from a management point of view. But because of its attractiveness, it appears useful to outline possible steps that would facilitate closer cooperation which might lead to a fully integrated solution at a later stage.

---

[14] See Feitelson and Haddad 1995.
[15] See Fisher 1995.
[16] GTZ 1996.

5.6 Steps Towards a Joint Approach in the Jordan Basin

A logical first step of collaboration among water management systems is joint water data monitoring because cooperating parties need reliable and mutually agreed data on water resources and water use and demand, regardless of the intensity of collaboration. The Israeli and Palestinian experts who worked on proposals for the joint management of shared aquifers also agreed on the priority need for data monitoring.

A second joint task would be the development of a common approach to waste water treatment and reuse, particularly in the Israeli and Palestinian territories which are linked in many ways. The Jordanian waste water system is also linked to the other territories since it discharges waste water into streams which are tributaries of the Jordan River. More broadly, environmental issues such as the disposal of wastes and hazardous materials, erosion control and sedimentation would benefit from closer collaboration. This is especially important because the quantity of waste water is rapidly increasing and treated waste water represents a major resource which has to be managed adequately to make it available in a useful form.

Since the water shortages in the Jordan Basin will require the development of additional water in the near future, Israelis, Jordanians and Palestinians have already started investigating the possibilities of joint efforts in this area, as mentioned above. Although no agreements have been reached, it is the declared intention of the three parties to continue jointly the work on the mobilization of additional water. Such work would include the preparation of interconnections between the various water systems that would allow the buying and selling of water among the parties and eventually the creation of a joint system.

Apart from the work on those technical issues, there are organizational and institutional questions which could be dealt with. The various ongoing activities in which the three parties are engaged bilaterally would preferably be handled by a joint commission or agency for the water sector. Such an organization could grow gradually, taking charge of tasks such as joint data monitoring, waste water issues and joint planning for the development of additional water resources. It could also explore the possibilities of privatization, the mobilization of additional water through privately built and operated desalination plants for the benefit of the three parties, and likewise the private construction of waste water treatment plants and conveyance facilities.

The development of jointly agreed standards for the water sector would also be a relatively easy task which would contribute to increased confidence in each other. Common standards for water monitoring, the delivery of drinking water, irrigation water, waste water treatment and discharge would be in the interest of all in a river basin shared by five peoples.

Other joint efforts could be training in water sector activities, the implementation of pilot projects and research on priority issues.

Today, the declared objectives of Israeli, Jordanian and Palestinian politicians are not very conducive to a joint approach in the water sector. They stress national interests, defend national development goals, solutions which they can control within national territories. They support subsidies to water users even if the lower water prices increase demand instead of encouraging water conservation. They advocate expansion of irrigation even if water is scarce and too expensive for the average farmer. They favour state agencies to be in charge of the water sector and hesitate to introduce reforms which could make the sector more efficient. This criticism of the political elite must be contrasted with a compliment to many Israeli, Jordanian and Palestinian water professionals who advocate a different approach.

In working groups, conference contributions and discussions, and independent publications, Israeli, Jordanian and Palestinian professionals, water managers, experts and academics have stressed the need for joint activities, the abolition of subsidies and the introduction of cost covering tariffs, a stop to irrigation expansion, much stronger conservation efforts, much more energetic waste water collection and treatment and comprehensive water management. The goals of these professionals are not identical but they point clearly towards a future in which an efficient, self-financing water sector will be in a position to satisfy water demand of the urban-industrial sector in the whole basin and support a water efficient agricultural sector.

In view of experiences elsewhere, it may be necessary to wait for emergencies such as droughts, environmental problems such as ground water pollution or failures of well fields until decision makers, pressured by the public, recognize the need for a change in policy. In the meantime, progress may be expected in other sectors like energy, where interconnection is being discussed, and telecommunications, where new technologies based on satellites will probably overtake traditional national networks. And further work on the joint management of shared water resources by professionals, such as the group dealing with shared aquifers, might show decision-makers that joint management and sharing of water resources is more advantageous than the traditional partition and separate management of water.

References

Deutsche Gesellschaft für Technische Zusammenarbeit GmbH (GTZ); Consulting Engineers Salzgitter GmbH (CES) 1996: Multilateral Working Group on Water Resources: Middle East Regional Study on Water Supply and Demand Management. Regional Overview Phase I (restricted distribution).

Feitelson, Eran; Haddad, Marwan 1995: Joint Management of Joint Aquifers, Jerusalem.

Fisher, Franklin 1995: The Economics of Water Dispute Resolution, Project Evaluation and Management: An Application to the Middle East, in: Water Resources Development, Vol.11, No.4.

Kolars, John 1994: Problems of International River Management: The Case of the Euphrates, in: Biswas, Asit K. (ed.): International Water of the Middle East, Oxford.

Lehn, Helmut; Steiner, Magdalene; Mohr, Hans 1996: Wasser - die elementare Ressource, Heidelberg.

Lonergan, Stephen C.; Brooks, David B. 1993: The Economic, Ecological and Geopolitical Dimensions of Water in Israel, Victoria, B.C., Canada.

Oslo B Interim Agreement between the State of Israel and the Palestinian Authority, Article 40: Water and Sewerage, September 18, 1995.
Radzio, Heiner (ed.) 1988: 1913 - 1988, 75 Jahre Ruhrverband/Ruhrtalsperrenverein, Gütersloh.
Rivera, Daniel 1996: Private Sector Participation in the Water and Waste Water Sector, The World Bank, Washington, D.C.
The World Bank 1993: Water Resources Management, A Policy Paper, Washington, D.C.
The World Bank 1994: A Strategy for Managing Water in the Middle East and North Africa, Washington, D.C.
Tilahun, Wondimneh 1979: Egypt's Imperial Aspirations over Lake Tana and the Blue Nile, Addis Ababa.
Treaty of Peace between the State of Israel and the Hashemite Kingdom of Jordan, October 24, 1994, Annex II, Water Related Matters.
United Nations Water Conference, Mar del Plata, Argentina, March 14 - 25, 1977.
Waterbury, John 1979: Hydropolitics of the Nile Valley, Syracuse, N.Y.

Part II
Regional Studies

6 The Jordan River Basin: Prospects for Cooperation Within the Middle East Peace Process?

Ines Dombrowsky

During the last fifty years, water has been one issue of dispute within the Israeli-Palestinian and Israeli-Arab conflict on political self-determination and sovereignty in the territory of 'historical Palestine'. Since the resumption of the Middle East peace process in 1991, water has been a focal point of the bilateral and multilateral peace talks. This contribution aims to explain the characteristics of this natural resource and its utilization in the Jordan River Basin, and to analyze achievements, further prospects and potential limits for regional cooperation on this crucial resource within the current Peace Process.

6.1 The Hydrological and Political Environment

Conflict over Jordan River water has been exacerbated by the overriding political conflict on one hand, and by the specific regional climatic, hydrological and geological natural conditions on the other. The territories of the riparians of the Jordan River, i.e. Lebanon, Syria, Jordan, Israel and the Palestinians in the West Bank, lie in a climatic transition zone of semi-arid to arid conditions. Rainfall undergoes high seasonal, annual and spatial variations and the geological and topographical formations drain rainfall across political boundaries. In fact, most surface and major ground water resources are transboundary, apart from the fact that the final political status of the Palestinian community and the definition of some borders within the region are still disputed. The Jordan River and its tributaries, which is the main surface water resource of the region, is shared by the five above mentioned riparians. Three major ground water basins, constituting the so-called West Bank Mountain Aquifer, are shared between Israelis and Palestinians (see Figure 6.1. and Table 6.1.).

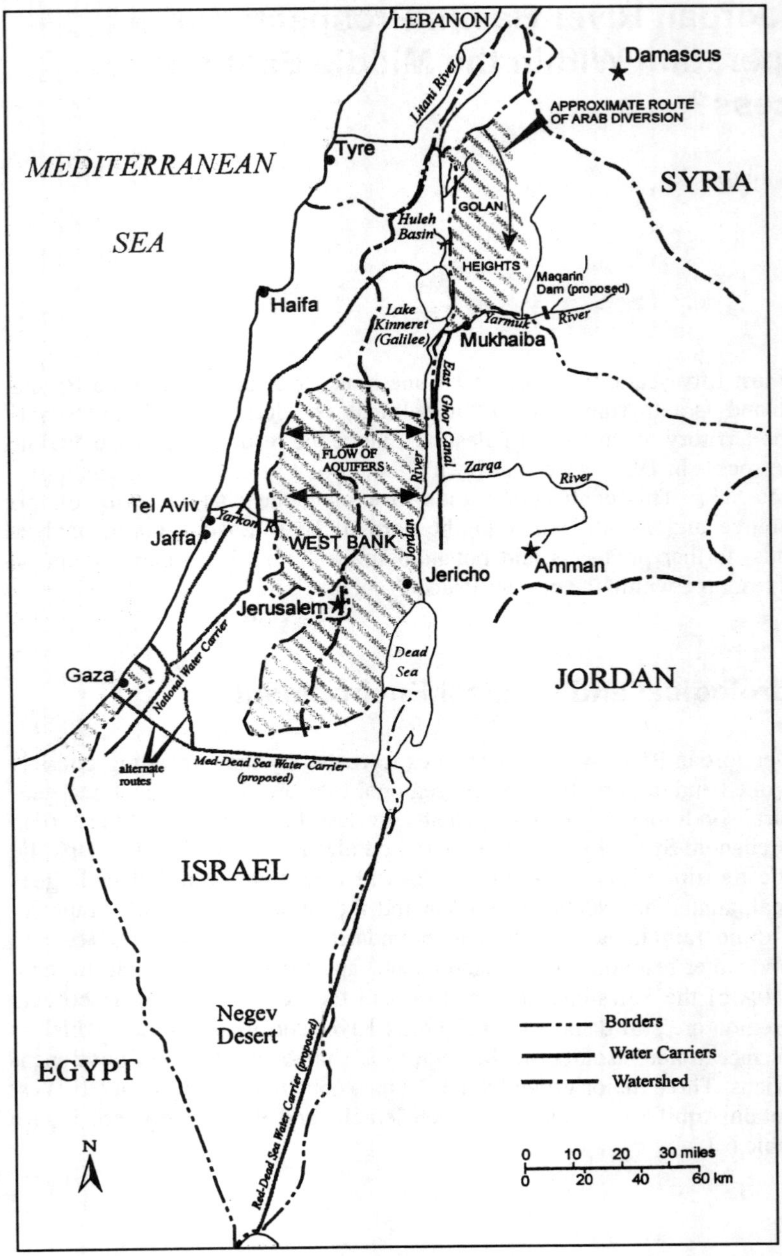

Figure 6.1. The Jordan Basin

While Lebanon and Syria have access to other major water catchment areas, the Israelis, Jordanians and the Palestinians mainly rely on the Jordan River's water and the ground water resources underlying their territories. The total renewable yield in the area of the three entities is limited and the theoretical average water availability per capita water is about 210 cubic metres per capita and annum (m^3/c/a). This might be sufficient for household and other municipal water demand, as well as the demand of light industries. However, it certainly is not enough for self-sufficiency in local food production by irrigated agriculture for which the average water need per capita has been estimated at 1,000 m^3/c/a under semi-arid conditions.[1] Moreover, demands are increasing due to rapidly growing populations and economic development, and the water quality is generally deteriorating.

In addition to the transboundary character and the limited total availability of the water resources, the geographic conditions require more sophisticated and energy-intensive water resource management than in many other regions of the world. This applies especially to the steep topographical gradients and differences in elevations, such as in the West Bank and on the Jordanian mountain ridges up to about 1000 metres high, the Jordan rift valley down to 400 metres below sea level, deep aquifers with wells which have been sunk up to 900 metres deep, and limited natural water reservoirs with Lake Tiberias as main surface water reservoir at 200 metres below sea level.

During the past decades, the implementation of major water development schemes has been one reason for conflict over water between Israel on one side and the Arab League on the other.[2] The four Arab-Israeli wars furthermore shaped the hydro-strategical positions of the riparians. All in all, the picture of regional water utilization patterns has been rather fragmented, since not only water resources utilization, but also water resources data have been disputed. One reason for this dispute is the fact that since 1967, the Palestinians had not only been prevented from the development of additional water resources in general, but also from gaining water resources data or having access to existing information. This situation improved to some extent in the context of the Peace talks. In the ongoing German sponsored Middle East Regional Study on Water Supply and Demand, executed by the Deutsche Gesellschaft für Technische Zusammenarbeit (GTZ) GmbH, Israeli, Jordanian and Palestinian study teams have for the first time established water balances on existing and scenarios on future water supply and demand within the region in a concerted action. The following information is based on data available within the existing agreements and on various other sources which have partly not been published yet.

---

1 Allan 1995.
2 Wolf and Ross 1992; Lowi 1993; Reguer 1993.

6.2 Present Utilization Patterns

Sources and supplies

The average annual amount of available Jordan and Yarmuk surface water has been estimated at 1320 million cubic metres per annum (MCM/a). Out of this, in 1994, Israel used about 650 MCM and Jordan about 350 MCM, whereas the Palestinians did not have access to these surface water resources. An estimated additional 200-250 MCM of Yarmuk water was used by Syria. The remaining surface water discharges into the Dead Sea.

Table 6.1. Current regional utilization of conventional water resources (MCM/a)

	Renewable water resources	**Israeli abstraction 1994**	**Palestinian abstraction 1994**	**Jordanian abstraction 1994**
Jordan River Basin	1,320	645	0	350 (incl. wadis)
Mountain Aquifer	679	483	118	-
Coastal Aquifer	240 (+55)	240	108	-
Other Aquifers in Israel	215 (+110)	283	-	-
Other Aquifers in Jordan	275 (+143)	-	-	420 (+87)
Total	2,784	1,651	226	857

Figures in brackets indicate additional fossil/non-renewable ground water resources
Sources: Oslo-II Agreement and Israeli, Palestinian and Jordanian data

The total amount of annual renewable ground water resources underlying the territories of Israel, Jordan, and the Palestinian territories has been estimated at 1460 MCM/a with varying water quality. In 1994, the total abstraction of ground water accounted for about 1740 MCM, about 1000 MCM by Israel, about 230 MCM by the Palestinians and about 510 MCM by Jordan. The balance shows an overdraft of about 280 MCM, and in fact some of the ground water basins in Jordan as well as the coastal aquifer in Israel and the Gaza Strip have been heavily overexploited in recent years. Such mining of non-renewable ground water stocks leads to dropping ground water tables and, depending on the circumstances, might result in the intrusion of deep saline water, negatively affecting the quality of the water. An overview of the current regional water allocation is indicated in Table 6.1.

In 1994, the total water supply in the three territories accounted for about 3040 MCM/a, including about 265 MCM/a of treated waste water. Out of this, 5.3 million Israelis used about 1900 MCM, 4.1 million Jordanians about 910 MCM, and 2.2 million Palestinians about 230 MCM (Figure 6.2.). Gross per-capita use was 360 m^3/c/a in Israel, 220 m^3/c/a in Jordan, and 110 m^3/c/a in the Palestinian territories (Figure 6.3.).

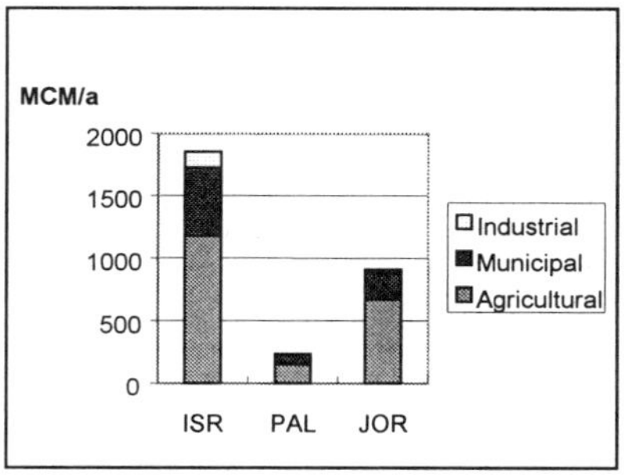

Figure 6.2. Regional total water use (1994)

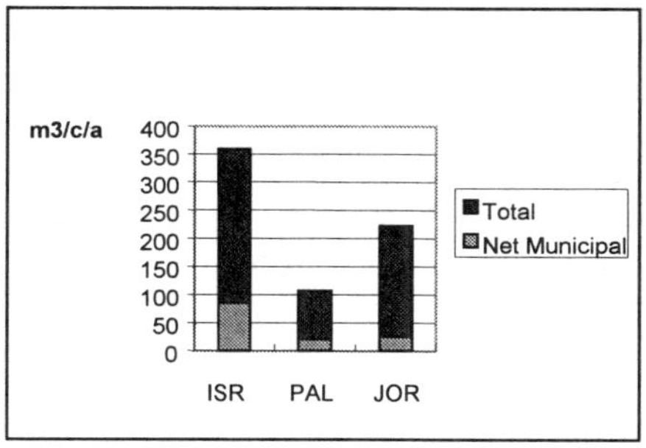

Figure 6.3. Regional per-capita use (1994)

Municipal Sectors

In the municipal sector, in 1994, net water use was about 230 litre per-capita per day (l/c/d) (or 85 $m^3/c/a$) in Israel, about 70 l/c/d (25 $m^3/c/a$) in Jordan, about 45 l/c/d (17 $m^3/c/a$) for Palestinians in the West Bank and about 75 l/c/d (28 $m^3/c/a$) in the Gaza Strip. In fact, gross municipal supply has been higher, and net municipal per-capita use figures are lower due to technical and administrative losses ('unaccounted-for-water') of about 15% in Israel, 55% in Jordan, 40% in the West Bank and 50% in the Gaza Strip.

The Israeli water supply system is highly integrated, and in good condition. Municipal tariffs are progressive, i.e. tariffs per unit increase with increasing total consumption, and cover costs. However, municipal per-capita water use is relatively high compared with international standards, and according to some Israeli experts up to 100-150 MCM/a could be saved in Israel through more determined efforts to manage urban water demand.[3] The water supply situation is different in Jordan and in the Palestinian territories, in which water supply infrastructure and the quality of the service often are in a comparatively poor state, with high unaccounted-for-water, water shortages in summer, and insufficient water supply and distribution infrastructure in rural areas. Municipal water tariffs in Jordan are progressive, but subsidized. However, average water use is comparatively low due to supply constraints. Water tariffs in the West Bank and the Gaza Strip so far are determined by municipal water agencies. Palestinians pay the highest municipal tariffs in the region since their water works are not subsidized at all. At the same time, their supply is much constrained.

Industrial Sectors

The industrial sector has played a minor role so far in the regional water balance, accounting for less than 7% of the water supply in all three entities and totalling about 130 MCM/a in Israel and 40 MCM/a in Jordan. Palestinian industrial use is not significant and is included in the data on municipal use.

---

[3] Arlosoroff 1995, p.4.

Agricultural Sectors

The largest water consumer in the region is agriculture, with an annual agricultural use of 1180 MCM in Israel (62% of supply, including 210 MCM of treated waste water), 150 MCM in the Palestinian territories (64% of supply) and 670 MCM (74% of supply, including 50 MCM of waste water) in Jordan. Agriculture today, as well as in the past, maintains a relatively high absolute level of water use in all three entities. Agricultural water supply is significantly subsidized in Israel and Jordan.

In Israel in recent decades efforts have been made to increase the productive (or technical) water efficiency of the agricultural sector through fixed allocations and the introduction of modern water-saving irrigation technologies. Drip and sprinkler irrigation is now applied on more than 90% of the irrigated areas. At the same time, Israel in recent years increased the supply of treated waste water to agriculture, which went along with a decline in agricultural fresh water use from about 1200 MCM/a in the year 1984/85[4] to 860 MCM/a in 1994. However, irrigation water tariffs are subsidized and the absolute level of use is still high compared with the total regional water availability (about 30%). At the same time, the contribution of Israeli agriculture to the Gross Domestic Product (GDP) steadily declined to only 2.4% in 1993. These figures indicate that although the productive water efficiency in Israel's irrigated agriculture is high, the allocative (or economic) water efficiency, the efficiency of the inter-sectoral water allocation, appears to be low.

In Jordan, agricultural water use in the Jordan Valley and in the Uplands has substantially increased in recent years. Irrigation technology is comparatively efficient in the Uplands, where farmers operate private ground water wells paying the full cost of water production, but the productive efficiency is especially low in the Jordan Valley and the southern Ghors. About 25% of the total irrigated area in Jordan uses low efficient surface irrigation methods. The contribution of agriculture to the GDP was 6% in 1993, and about 6% of the labour force is employed in agriculture.

In the West Bank and in the Gaza Strip, total agricultural use is significantly lower than in Israel and Jordan due to restricted access to water resources in the West Bank and to limited water availability in the Gaza Strip. Nevertheless, the contribution of agriculture to GDP and employment was still 31% and 33% respectively in 1992[5]. Drip irrigation methods account for about 60% of the total irrigated area.

Sanitation Sectors

The sanitation sector is best developed in Israel, where in 1994 an estimated total waste water of 320 MCM/a was collected and treated and about 210 MCM/a (32% of the total municipal and industrial supply) were reused. In Jordan, about 52% of the population is connected to waste water collection systems and about 50 MCM/a (24% of the total municipal and industrial supply) were reused in 1994.

---

[4] Schwarz 1990.

[5] Figures quoted in Elmusa 1996, p.23.

The fourteen existing treatment plants meet the desired standards only partly. In the West Bank and the Gaza Strip, about 50 to 70% of urban waste water is sewered, while the remaining urban and all rural waste water is disposed in septic tanks, percolation pits or even in open channels. Farmers along open channels use raw waste water for irrigation, but no treated waste water is reused since the efficiency of six out of seven existing treatment plants is very low or zero.

6.3 Comparison of the Hydro-strategic and Economic Positions

The above figures show sharp differences in the overall and in the per-capita water supply and use within the three entities, reflecting differences in the access to regional water resources and the hydro-strategic positions of the riparians, but also differences in the economic strength, as well as in development concepts and opportunities.

The Zionist movement and, after 1948, the State of Israel very early developed ambitious plans and concepts for systematic water resource development for the newly established Jewish Community. In the 1960's, despite substantial protests by the Arab League, Israel was able to implement the so-called National Water Carrier, diverting some 400 MCM/a of Jordan River water from Lake Tiberias to the coastal plain with the aim of integrating all available water resources into one water supply system, and allowing agricultural development in the Negev in the south, which has relatively fertile soils. In addition, Israel, as the lower riparian on the Mountain Aquifer, already 70 years ago started tapping this resource within the green line, over which it gained full control in 1967.[6] The Jewish community and later Israel from the beginning had a very strong vision of the need for systematic water resource development for the livelihood of the newly established community:

> Policy makers regarded the development of water resources ('the blood flowing through the arteries of the nation') as a vital mission of the development of the country.[7]

Although the specific perceptions of the role of the water may have changed in the meantime,[8] this philosophy has clearly shaped the physical state of the Israeli water sector until today.

After the occupation of the West Bank, the Gaza Strip, the Golan Heights and the Sinai by Israel in 1967, the Palestinians in the West Bank were cut off from the Jordan River, and water abstractions and use were regulated by the Israeli 'Civil Administration'. Palestinian total abstractions have been almost frozen at the 1967 level, despite a growing population and increasing demands.[9] In the Gaza Strip, regulation was not as strict, resulting in heavy over-abstractions of ground water

---

[6] Assaf et al. 1993, p.7.
[7] Galnoor 1980, p.159.
[8] Allan 1996.
[9] Dillman 1989.

due to the lack of alternatives. In both territories, coordinated water supply and sanitation infrastructure development was obstructed by the occupation power.

Jordan, as the lower riparian on the Jordan and on the Yarmuk rivers, has been restricted in its access to regional water resources. However, in 1961 it built the East-Ghor/King-Abdullah canal, diverting some 100 to 140 MCM/a of Yarmuk water into an irrigation canal parallel to the lower Jordan River. In the late 1970's, a World Bank coordinated lending programme for the construction of a long-planned Jordanian-Syrian dam project on the Yarmuk River, the Al-Wehda Dam, failed due to an Israeli veto against the project. In the framework of a joint protocol of the year 1987, Syria has increased its unilateral development of the upper Yarmuk River, and the long-term average water flows within Jordan are declining.

Today, all three riparians are already suffering from scarcity to a certain degree, but water scarcity is much more severe in Jordan and in the Palestinian territories, together with limited total and intermittent water supplies, and partly poor drinking water quality, especially in the Gaza Strip. At the same time, development options are limited in Jordan and Palestine, which have a much lower Gross National Product (GNP) per capita of US$ 1,120 and US$ 1,720 respectively compared to US$ 12,170 in Israel.[10] Their unfavourable hydro-strategic positions are even worse through the much lower economic capacity to invest in their water sectors.

The regional dispute on water rights is a conflict over the control of water resources and their allocation. The conflict has a political and ideological as well as an economic component. From an economic point of view, it is a conflict over access to relatively cheap fresh water resources. Conventional water resources in the region have been developed up to a marginal cost of about 20-30 cents/m^3 and any further development of additional conventional and or unconventional water resources steadily increases the marginal costs. Thus, changes in the control of the regional water resources would have an impact on total cost and on marginal unit water cost the entities have to bear in future. However, it is neither a conflict about the allocation of a 'free good', since conventional water resources have a cost, nor about the allocation of an absolutely limited good, since further fresh water production is at least theoretically possible through seawater desalination or water imports.[11] The distributional conflict on water is superimposed by the political and ideological components: water as a symbol of sovereignty and water as an instrument for other political goals.

6.4 Water in the Middle East Peace Process

The historical development of the control of the regional water resources, and the uneven control over and the access to regional water resources resulted in the Arab claim for a reasonable and equitable settlement of the water issue within the Middle East peace process. The architecture of the Middle East peace process that

[10] 1992 data, Elmusa 1996; World Tables 1994.

[11] There might be constraints in access to natural resources and capital to produce more water.

evolved after the Madrid Conference in October 1991, and the subsequent meeting in Moscow in April 1992, distinguishes bilateral and multilateral negotiation tracks. In the bilaterals, Israel separately negotiates with the Palestine Liberation Organization (PLO)/Palestinian Authority, Jordan and Syria all major political issues at stake. In the multilaterals, five separate working groups on specific topics have been established, with representatives of the region and other interested countries and international organisations addressing water, environment, economic relations, refugees and arms control. Syria and Lebanon have so far refused to participate in the multilaterals. According to the structure of the peace talks, the multilaterals are sub-ordinate and complementary to the bilaterals. All questions referring to international water resources and water rights are dealt with in the bilaterals, while on the multilateral level the issue of 'new and additional water' for the region has been put on the agenda. This distinction in the functions of the bilaterals and multilaterals has been an Israeli precondition for the negotiations, and, at least in the case of water, clearly structures the course of the talks and possible outcomes.

The Bilateral Track

All bilateral agreements that have been reached so far between Israel and Jordan and between Israel and the PLO address the water issue in one way or another. The most prominent water-related agreements are the so-called Annex II of the Israeli-Jordanian Peace Treaty of October 1994, and Article 40 of the Israeli-Palestinian Oslo-II Agreement of September 1995. The Israeli-Jordanian Annex II mainly refers to the utilization of the upper Jordan and Yarmuk river as well as to minor ground water resources in the Wadi Araba/Arava. Article 40 almost exclusively deals with the West Bank Mountain Aquifers, excluding the Jordan River and the Coastal Aquifer (Israel-Gaza). Both agreements basically maintain the existing allocations for Israel, and thus the principle of 'prior appropriation' and seek to develop so far unexploited water resources for the Jordanians and Palestinians. However, in both cases it is still not fully clear, which water resources the agreements exactly refer to, how they are to be developed and who will bear the costs of additional water resources development. These questions remain subject to further negotiations in bilateral water committees.

According to Annex II, Israel and Jordan aim to gain additional water for Jordan through better management of parts of the Jordan surface water system, including seasonal storage of Yarmuk water in Lake Tiberias, through additional storage systems at the Yarmuk and Lower Jordan rivers, and through the mobilization of other resources.[12] According to the agreement, a maximum of 225 MCM/a of additional water will be developed for Jordan,[13] including 50 MCM/a 'unidentified' water, for which the source has not been specified yet (Article I.3). Discussions on the implementation of the agreement are still ongoing two and a half years after the agreement was signed.[14] A main characteristic of Annex II is that two parties have bilaterally reached an agreement on major parts of the Jordan River, not consider-

[12] Annex II; Elmusa 1995; Libiszewski 1995.
[13] This figure is not stated in the agreement, but is the official Jordanian interpretation.
[14] E.g. The Star, 15-21 May 1997.

ing the claims of the three other riparians, i.e. Syria, Lebanon and the Palestinians in the West Bank. Their future participation in the utilization and management of the Jordan River is so far not specified.

In the case of the Mountain Aquifer, sharing benefits through joint development is even more complicated since mutual dependence and vulnerability of ground water resources is higher and prospects for win-win strategies through joint development schemes are smaller than in the case of surface water.[15] Article 40 of the Oslo-II Interim Agreement states that „Israel recognizes the Palestinian water rights in the West Bank. These will be negotiated in the permanent status negotiations and settled in the Permanent Status Agreement relating to the various water resources“ (Article 40.1). The current Israeli share will remain at 483 MCM/a, while the Palestinians may develop 70-80 MCM/a in addition to the current use of 118 MCM/a. Additional supply during the interim period is limited to 23.6 MCM/a for the West Bank and 5 MCM/a Gaza (Article 40.6 and Schedule 10). Development of additional water resources mainly refers to the so far lesser developed Eastern Aquifer (as well as 'other agreed resources' which are not specified in the agreement). The Eastern Aquifer is the least well understood of the three mountain aquifer basins. There are serious water quality problems (brackish water), and its exploitation is likely to be difficult and expensive. Although Israel is committed to bearing the capital cost for the development of 4.5 MCM/a of water, the agreement places a much larger financial and administrative burden on the Palestinians. This refers to further water resource development and to several stated principles with respect to ground water protection: „Preventing the deterioration of water quality...“, „Taking all necessary measures to prevent any harm to water resources, including those utilized by the other side“, „Treating, reusing or properly disposing of all domestic, urban, industrial and agricultural sewage“ (Article 40.3). Although the principles themselves are not disputable, the whole context of their application remains problematic, since rights and duties are unequally distributed. Israel is the main beneficiary of the use of the Mountain Aquifer through pumping in the confined zones in the area of the green line. The Palestinians, on the other hand, are responsible for the protection of the aquifer since they are in the main recharge area, where the carstic aquifer is unconfined and highly vulnerable. At the same time, the agreement does not say anything about obligations by the Israeli settlements in the West Bank.

It remains to be seen whether this unequal distribution of rights and duties will result in sustainable utilization patterns, environmentally and politically. It can be assumed that the incentive to adhere to certain agreements is much higher if these agreements are perceived as equitable. Even the 'Declaration of Principles' of September 1993, which paved the way for Palestinian-Israeli negotiations, stated that a 'Water Development Programme' would be „prepared by experts from both sides, which will also specify the mode of cooperation in the management of water resources in the West Bank and Gaza, and will include proposals for studies and plans on water rights of each party, as well as on the equitable utilization of joint

---

[15] Grey 1994.

water resources for implementation in and beyond the interim period".[16] No such studies have officially been conducted so far and it remains to be seen to which extent the two principles of equitable utilization and joint management will and can materialize in the final status agreement between Israelis and Palestinians.[17]

The third bilateral negotiation track refers to the negotiations between Israel and Syria, in which no agreement has been reached so far and which have officially been stalled since the Israeli elections in 1996. An Israeli-Syrian agreement is expected to be a precondition for any Israeli-Lebanese agreement.

The Multilateral Track

With the Multilateral Working Group on Water Resources (MWGWR), a forum has been established with 47 participating delegations to foster regional cooperation on certain issues such as data and new and additional water resources, and to coordinate donor activities. It was conceived as a self-implementing mechanism with regular meetings and reporting, which originally met biannually, and since 1995 annually. Since the meeting of the Arab League in March 1997, the future of the multilaterals is uncertain: due to the reluctance of the Israeli government to stop building new settlements in the neighbourhood of Jerusalem, the Arab League had recommended to its members to refrain from participating in all multilateral activities.[18]

The following topics are on the agenda of the Multilateral Working Group on Water Resources:

- Enhancing water data availability (Water Data Banks Project);
- Water management practices including conservation (Comparative Study on Water Institutions and Water Laws in the Middle East; several studies on management practices);
- Enhancing water supply (Middle East Regional Study on Water Supply and Demand Development; establishment of the Omani Desalination Research Centre);
- Concepts of regional water management and cooperation (Declaration of Principles for Cooperation; several training programmes).

The sub-ordination of the multilaterals to the bilaterals has been manifested in the concept of an artificial separation of 'waters', differentiating 'existing water resources', 'additional water resources developed pursuant to bilateral agreements' and 'new and additional water resources', which has been reflected in a Norwegian sponsored paper of the MWGWR, the Declaration of Principles for Cooperation on Water Related Matters and New and Additional Water Resources. The document states that "New and Additional Water Resources are only those potential water resources which are not Existing Water Resources, and which are not part of new and additional water resources developed pursuant to bilateral agreements (...). The development of New and Additional Water Resources will not adversely

[16] Annex III, Article 1.
[17] Elmusa 1996.
[18] The Economist, 5 April 1997.

affect the development or utilization of Existing Water Resources".[19] This concept clearly limits options for integrated regional water management and cooperation, since it systematically excludes the opportunity to reach efficiency gains through optimal management of all regional water resources.

Preliminary Conclusions

Within the peace process, the aspect of water scarcity and management has been broadly addressed, and various activities are underway which aim to improve the situation. However, despite major attempts by Jordan to implement Annex II and the activities of the Palestinian Water Authority in the framework of the Oslo II Agreement, there have been no significant changes with respect to the control over water resources. The bilateral agreements that have been reached so far basically maintain Israel's control over those 'existing' water resources which it had controlled before, and, in addition, allow Israel to control the development of additional water resources for the Jordanians and Palestinians pursuant to bilateral agreements, through the instrument of bilateral joint water committees. The existing water-related agreements so far reflect the distribution of power within the region and have been criticised, especially by Arab commentators, for lacking internationally favoured principles on the utilization of international watercourses. This is the case with respect to the principle of 'reasonable equitable utilization and participation', since the agreements manifest the principle of prior appropriation, not taking into account any other relevant factors listed in the ILA Helsinki Rules or ILC Convention on the Law of the Non-Navigational Uses of International Watercourses. With respect to cooperation, especially in the case of the Oslo II Agreement, they lack the principle of mutuality which, according to Article 8 of the ILC convention, should include the aspects of 'sovereign equality, territorial integrity, and mutual benefit'.[20] In line with bilateral agreements, concepts for an integrated and comprehensive joint management of all regional water resources are also broadly excluded from the agenda of the multilaterals. Under these circumstances, the question emerges of what the further potential for cooperation would be, and under which conditions it is likely that parties will agree on cooperative solutions and under which they will not.

6.5 Prospects for Further Regional Cooperation

Levels of Cooperation

An analysis of existing international water agreements shows three major levels of transboundary cooperation in order to avoid conflicts on shared water resources: sharing of information, division of water resources by volumes or guaranteed minimum flows and joint planning and management of water resources.[21] Sharing of information might be a first step to avoiding conflicts on water-related issues

[19] Part II, Articles 1.4, 2.1.
[20] See Elmusa 1996.
[21] Dellapenna 1995.

and to increasing the common understanding of the shared water resources system. This concept often goes along with the principle of prior notification to reach consent for a planned use. Prior notification, however, is only likely to lead to an agreement if enough water is available. But if water is scarce relative to demand, attempts to obtain consent to changes for water use may paralyze further development.

Agreements to divide the transboundary waters by volume aim to avoid conflicts through a restriction of the 'scope' of sovereignty. An apportionment of the resources by volume or a minimum flow guarantees the continued sovereignty of certain volumes and minimizes the need for further coordination. A division of water by volume, at least theoretically, may lead to solutions that are perceived to be equitable by all parties involved. However, in reality the division is more likely to reflect the distribution of power than normative principles. Major disadvantages of divisions by volumes are that they only coincidentally optimize the economic utilization of the resources, and that they become outdated when conditions change. Therefore, allocation agreements frequently require renegotiations, which are at minimum time and money consuming, but similar to the principle of prior notification, they are likely to reconfirm the current utilization while preventing the adaptation to changing conditions and further development of so far less developed riparians.

These problems might be prevented and additional efficiency gains be achieved, if the planning and management of the shared water resource is jointly undertaken. Examples of joint water management are rare. This suggests that nations are seldom willing to compromise their sovereignty on water resources or submit their futures to international institutions to plan, and possibly also to construct and operate water supply system.[22] However, there may be three reasons for such comprehensive cooperation: economic gains, environmental sustainability and political stability.

Economic Reasons for Cooperation

In the Jordan River Basin, due to the spatial and temporarily uneven distribution of supplies and demands, and due to the topography as well as the unequal development of water supply systems, it is very likely that a significant potential for efficiency gains through the optimized management of the regional water resources exists. Economic efficiency gains could be achieved through lower costs on the supply side and through higher benefits on the demand side. These saved costs or gained benefits could be shared by the parties and could be incentives for joint action (win-win-strategies).

First of all, there are economic reasons against an armed conflict over shared water resources in the Jordan River basin. The Harvard Middle East Water Project, a joint effort by Israeli, Palestinian, Jordanian and North American scholars, has estimated the value of the water in dispute, i.e. the Upper Jordan River, the Yarmuk River and the Mountain aquifer, at a current maximum of about US$ 110 million per year, rising to a maximum of less than US$ 500 million per year in the

---

[22] Dellapenna 1995, pp.286.

year 2020 (in 1990 dollars).[23] If one compares the estimates to the GDPs of US$ 60.594 billion (1993) in Israel, US$ 3.662 billion (1993) in Jordan and US$ 3.435 billion (1992) in the West Bank and Gaza,[24] it becomes clear that the actual value of the water in dispute is about 0.2% of the Israeli GDP and about 3% of the Jordanian and Palestinian GDPs. This might suggest that it is not economically viable to fight over water resources of the region. The value of the disputed water also indicates the order of magnitude of the potential benefits from cooperation on existing water resources, which appear to be limited. However, these benefits also have to be compared to the opportunity cost of non-cooperation, which might be significantly higher if the demand for fresh water were met only through the mobilization of new and additional water resources, such as seawater desalination or water imports.

Notwithstanding the mobilization of new and additional water, in all three entities, a further potential for optimized management of the indigenous water resources through internal measures exists. These include the enforcement of water demand management concepts, the limited additional development of surface and ground water resources, the rehabilitation and extension of the existing water supply and distribution systems, the extension of waste water treatment and reuse, the substitution of fresh water through treated waste water in irrigation and the intersectoral reallocation of fresh water to higher value uses, and, to a limited extent, brackish water desalination. Generally spoken, most of these internal measures are economically advantageous compared to the mobilization of new water. In addition, many of the infrastructure measures listed above would be a necessary condition for the mobilization of new water. Otherwise expensive desalinated water might be lost through leakage or might pollute existing ground water resources after its utilization.

Not all of the suggested internal measures are currently politically accepted. In fact, measures related to the technical or productive efficiency of the use of water are more easily implemented than those advocating economic or allocative efficiency (reallocation). A large number of activities referring to technical efficiency are currently either advocated by representatives of the water sectors of the region or already under implementation. However, an additional potential appears to exist with regard to an optimization of the economic or allocative efficiency of the use of water, but such concepts require major political changes and would have socio-economic impacts.

Taking this into account, the order of magnitude and the size of future water deficits in the entities to a certain extent will depend on future internal water sector policies and utilization pattern. Under the assumption that all theoretically feasible internal measures will be realized, and mainly considering domestic and industrial fresh water needs, the following rough picture can be drawn:

[23] Fisher 1995.

[24] GTZ 1996; Elmusa 1996.

- In Israel, it is likely that domestic and industrial demand can largely be covered through existing resources in the long-term, and, depending on the degree of fresh water substitution in agriculture, there might even be a surplus of fresh water.
- In Jordan, an immediate deficit of some 100 MCM/a exists which, despite increasing municipal demand, could be reduced over time if the whole range of internal measures mentioned above would be implemented. Still, it is likely that a deficit would remain.
- In the West Bank, the Oslo II Agreement allows for limited additional water development which might cover the immediate domestic and industrial needs. However, it is likely that the additional allocation specified in the interim agreement will not be sufficient in the medium to long-term, calling for at least an additional 100 MCM/a.
- In the Gaza Strip, due to the existing ground water overdraft, there is an immediate deficit of about 50-100 MCM/a. Since the potential for internal alternatives is limited and demand is expected to rise, the deficit is likely to continue and will possibly increase to about 200 MCM/a in the long-term.

This scenario indicates that within the coming two decades the domestic, municipal and industrial needs of the region could largely be met through indigenous water resources, if the right infrastructure were put into place, if the irrigation sector primarily relied on treated waste water, and if those sub-areas that already today are short of water, e.g. Greater Amman, the Gaza Strip and the northern West Bank were connected to the existing water systems, i.e. if some of the fresh water in the region were reallocated. In the medium to long-term, it is likely that at least limited amounts of new and additional water will have to be developed to satisfy the domestic and industrial fresh water demands. If no intra-regional reallocation of fresh water emerges, the need for new and additional water will be higher.

This implies that the order of magnitude and the schedule for the mobilization of new and additional water to meet the currently unsatisfied and the long-term incremental water demands in the region will not only depend on future internal water policies, but also on future modes of international cooperation. The need for further investment in new and additional water sources would definitely be higher if no water were reallocated, and if existing and future deficits had to be bridged by new water only.

Cost estimates for various sub-alternatives for the mobilization of new water indicate that under current political and technological conditions, least cost solutions would be seawater desalination by reverse osmosis (RO), the so called 'intersea' project, connecting either the Mediterranean or the Red Sea and the Dead Sea, and using the hydrostatic pressure differential between the sea and the Dead Sea level for seawater desalination by reverse osmosis, or water import from the Manavgat River in Turkey to the Mediterranean coast by used tankers. Water import from the Litani River in Lebanon would be comparatively cheap but is not considered politically feasible under current conditions. Water import from the Manavgat River in Turkey by new tankers or water import from the Ceyhan and Seyhan rivers in Turkey or from the Euphrates River in Iraq by pipeline in any case would be significantly more expensive, and the latter two options are politically very complex. Water import by vinyl bags is expected to be cheap, but has not been technically proven yet. A comparison of unit cost estimates discounted at 5% interest rate (1996 prices) at different demand centres in the region is presented in Table 6.2.

New water would thus be primarily available on the seashore or at the Dead Sea. Seashore desalination is the only option that could be implemented unilaterally by Israel on the Mediterranean or Red Sea, by the Palestinians in the Gaza Strip or by Jordan in Aqaba on the Red Sea. For the West Bank, which is landlocked, cooperation in every case would be required for the mobilization of new water. All import options as well as the intersea project would require certain modes of cooperation.

Comparing the various alternatives for new water, it appears that at least for some sub-regions, projects relying on cooperation could be cheaper than unilateral projects. As mentioned above, the only option for Jordan unilaterally to mobilize new and additional water is sea water desalination at Aqaba and conveyance to the Amman region. Preliminary cost estimates indicate that this option would be relatively expensive and that it could be cheaper either to import water from Lebanon, to desalinate sea water at Haifa and to convey it to Amman, or to participate in one of the intersea alternatives, all requiring cooperation.

Further optimization might be possible through a trade-off between new and existing water resources. For example, it seems worthwhile to check whether part of the transport costs from the Mediterranean coast to Amman could be saved if Jordan River water were supplied to Amman, instead of transporting Jordan River water from Lake Tiberias through the existing Israeli water carrier to the coast and desalinated water from the coast to the Jordan Rift Valley. An interconnection of new and existing systems could also be advantageous with respect to the important inter-seasonal and inter-annual regulation of the water supply on a regional scale, making use of the existing systems and reservoirs to the greatest extent possible.[25]

It can be concluded that the future degree and modes of cooperation will be one major factor for the total water system costs of the region as a whole, and for the respective unit costs at which water could be supplied to the various sub-regions. However, the magnitude of the gains of cooperation depends on many factors and has not yet been conclusively determined. Therefore, it remains open whether potential economic gains might be a decisive incentive for regional cooperation. In any case, the economic argument will only be a necessary but not sufficient condition for cooperation, and levels and modes of cooperation will primarily depend on the overall political situation.

---

[25] E.g. Küffner 1992.

Table 6.2. Alternatives for mobilization of new and additional water

Option	Sub-option	Quantity (MCM/a)	Unit water cost discounted at 5% (US$/m³) TLV	GAZA	RAM	AMM
Sea water	RO desalination plant	50	0.70	0.70	0.80	1.00
Desalination	Med-Dead intersea scheme	800	n.a.	n.a.	0.70	0.70
	Red-Dead intersea scheme	850	n.a.	n.a.	1.00	1.00
Water import	Used tankers	200	0.90	0.90	1.00	1.10
by Sea	New water tankers	200	1.10	1.10	1.30	1.40
(Manavgat R.)	Large Vinyl bags	200	0.60	0.60	0.70	0.80
Water import	Pipeline from Seyhan/Ceyhan R.	150	2.00	2.20	n.a.	1.70
by land	Pipeline from Euphrates R.	150	1.50	1.60	n.a.	1.10
	Pipeline from Litani R.	150	0.80	1.20*	n.a.	0.70

TLV = Tel Aviv; GAZA = Gaza City; RAM = Ramallah; AMM = Amman
* Via West Bank
Source: Middle East Regional Study on Water Supply and Demand Development, Rounded Figures

Political and Environmental Reasons for Cooperation
Besides potential economic benefits, joint management of shared water resources could be environmentally and politically advantageous. Treating the shared resources as a unitary resource is more likely to satisfy environmental concerns, since extractions and use are jointly controlled, and since the flexibility in dealing with droughts could be enhanced. Management concepts based on mutuality are furthermore more likely to lead to sustainable utilization patterns, since the parties have greater incentives to protect the resource and abide by their duties when rights and duties are equitably distributed. Such an equitable distribution of rights and duties does not necessarily require a joint management system, but it could also be realized through an equitable division of water by volume. On the other hand, the difficult question of an equitable apportionment of water rights could be avoided if the water resources were jointly managed. In this case, the distribution question would probably shift to the question of the division of costs for joint development projects.

Although cooperation on shared water resources and possibly joint management is appealing from an economic, environmental and a political point of view, there are a number of political constraints as well.

6.6 Limits of Cooperation

International Arena
The overriding political constraint to a more equitable division of water resources and a joint management approach in the Jordan River Basin at the present stage is the so far undefined final status of the Palestinian community and its territorial

control, and the pending peace between Israel on one side and Syria and Lebanon on the other. The Oslo II Agreement and the planned expansion of Israeli settlements in the West Bank show a tendency towards long-term geographical control of the West Bank. Members of the Israeli government use the question of the control of water resources as an argument against a total withdrawal from the West Bank,[26] arguing that water allocation to the Palestinians could even be augmented as long as Israel maintains control over the water resources. This ideology aims to legitimize the continued occupation, despite the fact that ground water resources in the West Bank and Israel are shared. Water in this case is clearly instrumentalized. However, not only the Israeli, but also the Palestinian side perceives control over water resources as matter of sovereignty. In the latter case, this is in so far not surprising since the occupation imposed heavy restrictions in the water sector, and since the Palestinian Authority is dependent on a certain volume of water resources for economic development. The latter argument does not necessarily need to be equated with sovereignty. However, it may be argued that the Palestinian Authority needs to establish its rights and its own institutional and managerial capacities before it can enter into a joint management mechanism as an equal partner.

When it comes to the import of water, although cooperation would be based on an exchange relation (money-water), the party relying on water resources which are controlled by other parties might object to being dependent on such water resources.

The perceptions of water as a political matter of sovereignty on one hand, and as an economic good on the other are not easily reconciled in a coherent approach. However, cooperation that would involve all water resources (existing and new), and in which all entities involved would give up sovereignty over their current control over regional water resources, could lead to a situation of mutual dependency and balance, which might be more acceptable for all parties concerned.

These considerations remain theoretical, as long as the overriding political atmosphere lacks a certain degree of trust and confidence.

National Arenas

Political constraints to regional cooperation in the Jordan River Basin are not only related to problems in international relations, but also to the reluctance of stakeholders in the water sectors of Israel and of Jordan to give up vested interests and apply economic principles as a prerequisite for a rational regional approach. In both countries, historically the water sector was closely connected to the agricultural sector, which inter alia resulted in substantial subsidies to water consumed in agriculture. Drastic changes in the price of water supplied to agriculture would have substantial repercussion on the agricultural sectors within these countries, and at minimum would make the bulk of existing agriculture uneconomic. Nevertheless, internal discussions and processes are ongoing and step by step changes and transitions in the water sectors can be observed. Allan states:

> The rates at which perceptions are changing differ widely among the numerous interest groups involved (...). It is the lack of symmetry between the knowledge and expectations

[26] Frankfurter Allgemeine Zeitung, 22 May 1997.

> of the technological and political leaderships of the water-sharing countries and the assumptions and expectations of the major water-using constituency, irrigated farming, that has already caused significant problems for those addressing water issues and particularly water sharing (...). Both national and sub-national groups with a vital interest in water are emerging from a period when only a tiny minority of technocrats and scientists were prepared to view water as an economic rather than a non-substitutable strategic resource, to a new period when an increasing number of such officials and politicians are prepared to take a perspective that gives priority to the overall economy and not just to the arguments of those whose livelihoods depend on the use of large volumes of water supplied free or nearly free.[27]

In Israel, a report has been recently published which suggests drastic changes in the Israeli water sector, calling for the introduction of private sector competition at the level of water utilities, levies on water production and increased charges for the agriculture sector, while offering alternatives to farmers. While the report is supported by the Ministry of Finance, its recommendations have been rejected by the Water Commissioner and the Ministries of Agriculture and of Infrastructure.[28] In Jordan, a new water strategy has been approved by the parliament recently. At the same time, the Minister of Water and Irrigation announced a stricter control of illegal wells and new prices for the municipal sector to be effective from October 1997, which will lead to higher accountability of the Water Authority of Jordan.[29] These examples show that internal discussions are ongoing on different levels, but transitions at the policy level may take some time.

Economic Inequalities

Even if the concept of water as an economic good is accepted by all parties, the uneven economic situation of the riparians, and the uneven state of infrastructure may still lead to difficulties with respect to joint management approaches. Who would have to bear the investment cost for internal infrastructure measures? If water prices within the whole region were based on the concept of marginal cost pricing, the unequal economic ability to pay for water might still lead to political and social tensions. On the other hand, water charges of above US$ 1/m^3 are becoming increasingly common in domestic and industrial sectors of the region. In the domestic sectors, for reasons of social equity, a cross-subsidization will be needed. In agriculture, tariffs above US$ 1/m^3 at least could be paid for special agriculture crops, off-season and intensive production. Concepts and solutions are needed for traditional farming, which relies heavily on subsidized water prices. In any case, it is likely that total system costs would be lower with regional cooperation than without. Nevertheless, economic development and adjustment in the level of welfare will be crucial for joint management. In the short-term, it should be recognized that development options for Jordan and the Palestinians will also depend on the endowment with water resources and improvements in their situation.

---

[27] Allan 1995, p.52.

[28] The Jerusalem Post, 19 May 1997.

[29] The Star, 15-21 May 1997.

6.7
The Role of the International Community

In the current process, the international community is playing a dual role: to finance and support certain activities, but at the same to coordinate, mediate, and facilitate these activities.

So far, it seems that the parties involved tend rather to approach the international (donor) community for support in financing 'grand solutions' than to support joint management mechanisms. Such 'grand solutions' may include the Red-Dead-Sea-Canal which is currently being studied at the prefeasibility level by the Harza Group. However, other studies underway demonstrate that it is rational to invest in existing systems first. Even if one concluded that the mobilization of new water were necessary in the medium-term, the implementation of 'grand solutions' at the present stage would appear to be premature and a disincentive for a comprehensive approach towards regional cooperation including all water resources. Therefore, the role of the international community is very delicate. On one side it is expected to contribute economic support to the needy societies in their current difficulties, but on the other side it should give incentives and foster solutions that are sustainable in the long run. Ongoing projects show that external support, coordination and mediation is important. The support of joint management mechanisms could be a major contribution in future.

6.8
Conclusions

Due to hydrological and socio-demographic conditions, water resource management is extremely challenging in the Jordan River Basin, and will be even more so in future. This situation is aggravated by the general political conflict which is interconnected with the specific water conflict and the uneven control over, and the allocation of, the regional water resources. So far, the problem with respect to water in the region is twofold: the general water management challenge, and the political conflict on water sovereignty. Within the ongoing peace process, water issues have been addressed, and the Israeli-Jordanian and Israeli-Palestinian agreements foresee additional water development for Jordan and the Palestinians. A final agreement between Israel and the PLO, and peace between Israel on one side and Syria and Lebanon on the other are still pending.

The bilateral agreements reached so far, as well as the activities on the multilateral level, are contributing to an improvement, but are still far away from solutions that would lead to a more equitable and economically optimized allocation of the regional water resources. The political wish to maintain absolute control over water resources for, at least partly, ideological reasons is a substantial obstacle to joint management in which all parties would give up sovereignty on water to a joint body. Joint management on the other hand could lead to a minimization of the total system costs, to better protection of water resources including drought management and could enhance regional stability. However, such a concept remains far from realization, since it requires further changes in the national water sectors of the region, as well as solutions against socio-economic imbalances. The water question in the region is therefore directly connected to the economic situation as a whole.

Regional cooperation on water remains a complex notion. Real transparency and exchange of data and information may be a first step to narrowing the diverging perceptions of the problem and contributing to more rational water resource planning in the region.

References

Allan, Tony 1995: A Transition in the Political Economy of Water and the Environment in Israel-Palestine, in: Marwan Haddad and Eran Feitelson: Joint Management of Shared Aquifers: The Second Workshop. Nov. 27-Dec. 1, 1994.

Allan, Tony 1996: The Political Economy of Water: Reasons for Optimism but Long Term Caution, in: J.A. Allan (ed.): Water, Peace and the Middle East. Negotiating Resources in the Jordan Basin. London, New York.

Arlosoroff, Shaul 1995: Water Resources Management within Regional Cooperation in the Middle East, in: Marwan Haddad and Eran Feitelson: Joint Management of Shared Aquifers: The Second Workshop. Nov. 27-Dec. 1, 1994.

Assaf, Karen; Nader Al Khatib; Elisha Kally; Hillel Shuval 1993: A Proposal for the Development of a Regional Water Master Plan. Israel/Palestine Center for Research and Information, October.

Declaration of Principles For Cooperation on Water-related Matters and New and Additional Water Resources 1996. Joint Statement by the State of Israel, the Palestinian Authority and the Hashemite Kingdom of Jordan. Oslo, June 12.

Dellapenna, Joseph 1995: Designing the Legal Structures of Water Management Needed to Implement the Israeli-Palestinian Declaration of Principles, in: Marwan Haddad; Eran Feitelson (eds.): Joint Management of Shared Aquifers: The Second Workshop. Nov. 27-Dec. 1, 1994.

Dillman, Jeremy D. 1989: Water Rights in the Occupied Territories, in: Journal of Palestine Studies, Vol.19, No.1.

Elmusa, Sharif 1995: The Jordan-Israel Water Agreement: A Model or an Exception?, in: Journal of Palestine Studies Vol.24, No.3.

Elmusa, Sharif 1996: Negotiating Water: Israel and the Palestinians. Institute for Palestine Studies, Washington, D.C.

Frankfurter Allgemeine Zeitung: Todesstrafe für Landverkauf, 22 May 1997.

Fisher, Franklin 1995: The Economics of Water Dispute Resolution, Project Evaluation and Management: An Application to the Middle East, in: Water Resources Development, Vol.11, No.4.

Galnoor, Itzhak 1980: Water Planning: Who Gets the Last Drop?, in: R. Bilski et al. (eds.): Can Planning Replace Politics? The Israeli Experience. Den Haag.

Grey, David 1994: The Development of the Water Resources of the Occupied Palestinian Territories: Some Key Issues, in: Isaac, Jad; Hillel Shuval (eds.): Water and Peace in the Middle East. Amsterdam: Elsevier.

GTZ 1996, Deutsche Gesellschaft für Technische Zusammenarbeit GmbH; Consulting Engineers Salzgitter (GmbH) CES: Multilateral Working Group on Water Resources: Middle East Regional Study on Water Supply and Demand Development. Regional Overview Phase I, August (restricted distribution).

GTZ 1997, Deutsche Gesellschaft für Technische Zusammenarbeit GmbH; Consulting Engineers Salzgitter (GmbH) CES: Multilateral Working Group on Water Resources: Middle East Regional Study on Water Supply and Demand Development. Regional Overview Phase II (under preparation).

ILC 1997: International Law Commission: Convention on the Law of the Non-Navigational Uses of International Watercourses, reproduced in: International Legal Materials, 3/1997, pp.700-720.

Küffner, Ulrich 1992: Water Transfer and Distribution Schemes, in: Water International, 18 (1).

Libiszewski, Stephan 1995: Das Wasser im Nahostfriedensprozeß - Konfliktstrukturen und bisheriges Vertragswerk unter wasserpolitischer Perspektive, in: Orient Vol.36, No.4.

Lowi, Miriam 1993: Water and Power. The Politics of a Scarce Resources in the Jordan River Basin. Cambridge.

Oslo II Interim Agreement Between the State of Israel and the Palestinian Authority. Article 40: Water and Sewerage, September 28, 1995.

Reguer, Sarah 1993: Controversial Waters: Exploitation of the Jordan River, 1950-80, in: Middle Eastern Studies, Vol.29, No.1.

Schwarz, Joshua 1990: Management of Water Resources in Israel, in: Israel Journal of Earth Sciences, Vol.39.

The Jerusalem Post: Water Commission. Don't Break up Mekorot Monopoly, May 19, 1997.

The Star: Government Strikes Hard to Save Country's Water. Amman, Vol.7, No.51, 15-21 May 1997.

Treaty of Peace Between The State of Israel and The Hashemite Kingdom of Jordan, October 24, 1994: Annex II: Water Related Matters.

Wolf, Aaron; Ross; John 1992: The Impact of Scarce Water Resources on the Arab-Israeli Conflict, in: Natural Resources Journal, Vol.32, No.4.

World Tables 1994. International Bank for Recontruction and Development, Washington, D.C.

7 Conflicts on the Euphrates: An Analysis of Water and Non-water Issues

Waltina Scheumann

Since the beginning of the 1980's, several events have triggered headlines in the international and the German press upon the presumption that water scarcity will lead to armed conflicts over the allocation of shared rivers. The Euphrates-Tigris River Basin (Figure 7.1.), shared by the co-riparian states Turkey, Syria and Iraq, is supposed to be one place in the world where war could break out. It is without doubt that with rising population and the associated increase of water demands, the allocation of shared rivers generates a high potential for conflicts. But the equation 'water scarcity' equals 'war' is somehow dangerous, as it neglects the states' efforts to settle conflicts. In the last hundred years, numerous river treaties have been signed which show that, even under difficult conditions, solutions can be achieved. Whether cooperation is forthcoming or not depends not only on the water issue. In almost all cases, cooperation was positively influenced by outstanding issues, i.e. not directly related to the allocation of the shared rivers, and this holds true for conflicts as well. It is important to note that, historically, the negotiation process over the Euphrates River has shown phases of conflicts with changing coalitions, and agreements have been reached to some extent. The present pattern of conflict derives not only from the disputed allocation of the Euphrates' water but from outstanding policies, which instrumentalize, at least in some respect, the unresolved water issue.

The main areas which are covered here are to what degree the conflicts are related to water scarcity; the process of negotiation with its agreements; the conflicting issues over the allocation of the Euphrates' water; the role which non-water issues play, and the relevance of not yet negotiated topics for future agreements.

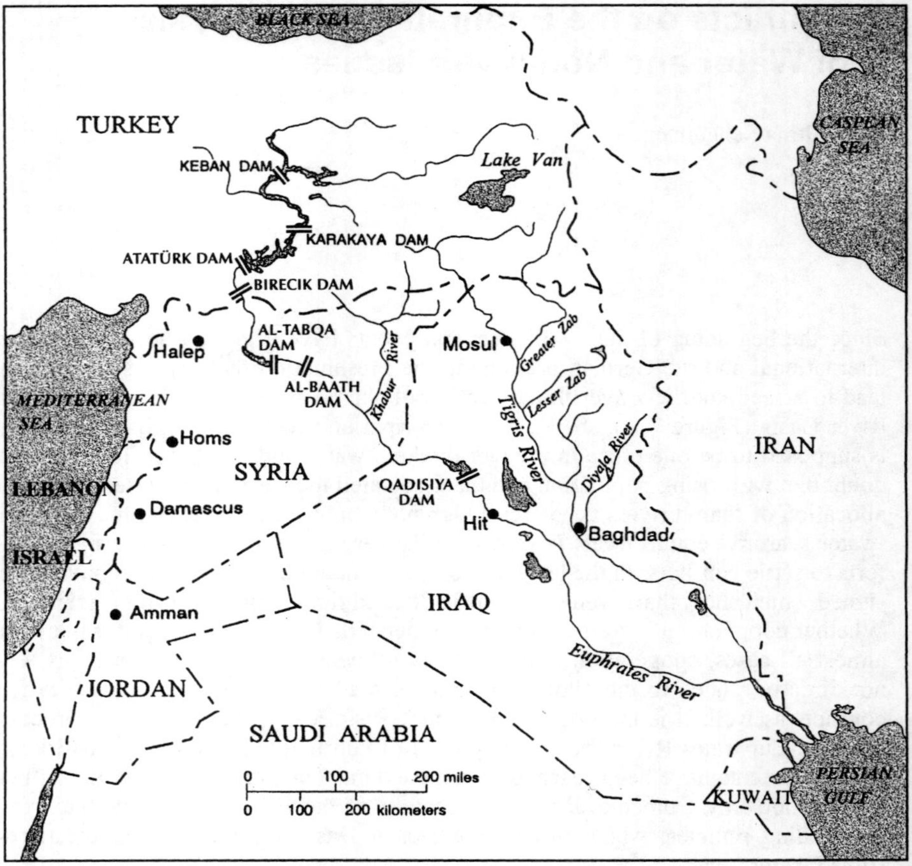

Figure 7.1. The Euphrates-Tigris Basin

7.1
Unilateral Development of Irrigated Agriculture

The co-riparian states are unilaterally strengthening their efforts to develop water resources, which will increase their hydropower potentials and extend irrigated agriculture. However, it is their development plans for irrigated agriculture which pose the main threat to water allocation, as agriculture does, and will, consume by far the biggest share of the Euphrates' waters.[1] The main impact, which will change the patterns of water demand between the three states, comes from the huge

[1] On country-wide levels, in Syria 83%, in Iraq 92% and in Turkey more than 50% of the water already developed are used in agriculture.

Turkish Southeastern Anatolian Project, known as the GAP.[2] This integrated development programme focuses on agricultural and industrial issues in seven southeastern provinces of Turkey. As Turkey suffered badly from the two oil crises during the 1970's, it started the exploitation of indigenous energy resources to limit dependency on imported oil. Within the boundaries of the GAP, hydroelectrical power plants can provide 25% of Turkey's electricity demand, by producing 85% of its hydroelectrical energy. In this less-developed region there is a high potential of irrigable land, i.e. 40% of Turkey's arable land. Within the GAP it is intended to open up about 1.091 million ha to be irrigated from the Euphrates (see Table 7.1.) and another 600,000 ha from the Tigris.

Table 7.1. Stages of Turkish irrigation projects on the Euphrates[3]

Source of irrigation water	Area (ha)	Stage
Atatürk Dam (1990/91)	706,281	40,000 operated
Suruc-Yaylak Project	146,500	reconnaissance
Camgazi Dam	6,536	construction
Gömikan Dam	7,762	Masterplan
Kocali Dam	21,605	Masterplan
Büyükcay Dam	12,322	Masterplan
pumped from Atatürk reservoir	29,599	Masterplan
Cataltepe Dam	71,598	feasibility study
Hancagiz Dam	7,330	in operation
Kayacik Dam	13,680	construction
Kemlin Dam	1,969	feasibility study
Birecik Dam	66,021	feasibility study
Total	1,091,203	

Not suprisingly, it is this GAP and, in particular, the Atatürk Dam which has been seen as a threat to the Arab countries Syria and Iraq. Turkey's upstream position in the river basin would enable it to cut off the river flow, and with the Turkish projects being implemented, the Euphrates' flow would significantly be reduced. The situation is worsened by the fact that Turkey has shown, thus far, the financial and technological ability to construct huge hydraulic infrastructures, i.e. the Atatürk Dam and the Urfa Tunnel, which diverts water from the reservoir to the Urfa-Harran and the Mardin-Ceylanpinar plains[4]. However, in 1997, estimates consider that about 140,000 ha are irrigated from the Euphrates in Turkey, including 40,000

[2] GAP is the Turkish abbreviation for 'Güneydogu Anadolu Projesi', i.e. the Southeastern Anatolian Project.

[3] From Ünver 1997, p.199.

[4] The two Urfa tunnels each of which has a diameter of 7.62 m, have to overcome a distance of 26.4 km from the Atatürk Dam reservoir since they reach the city of Sanliurfa where water is diverted for the two plains.

ha which are irrigable as a result of the constructed GAP infrastructures. (see Figure 7.2.)

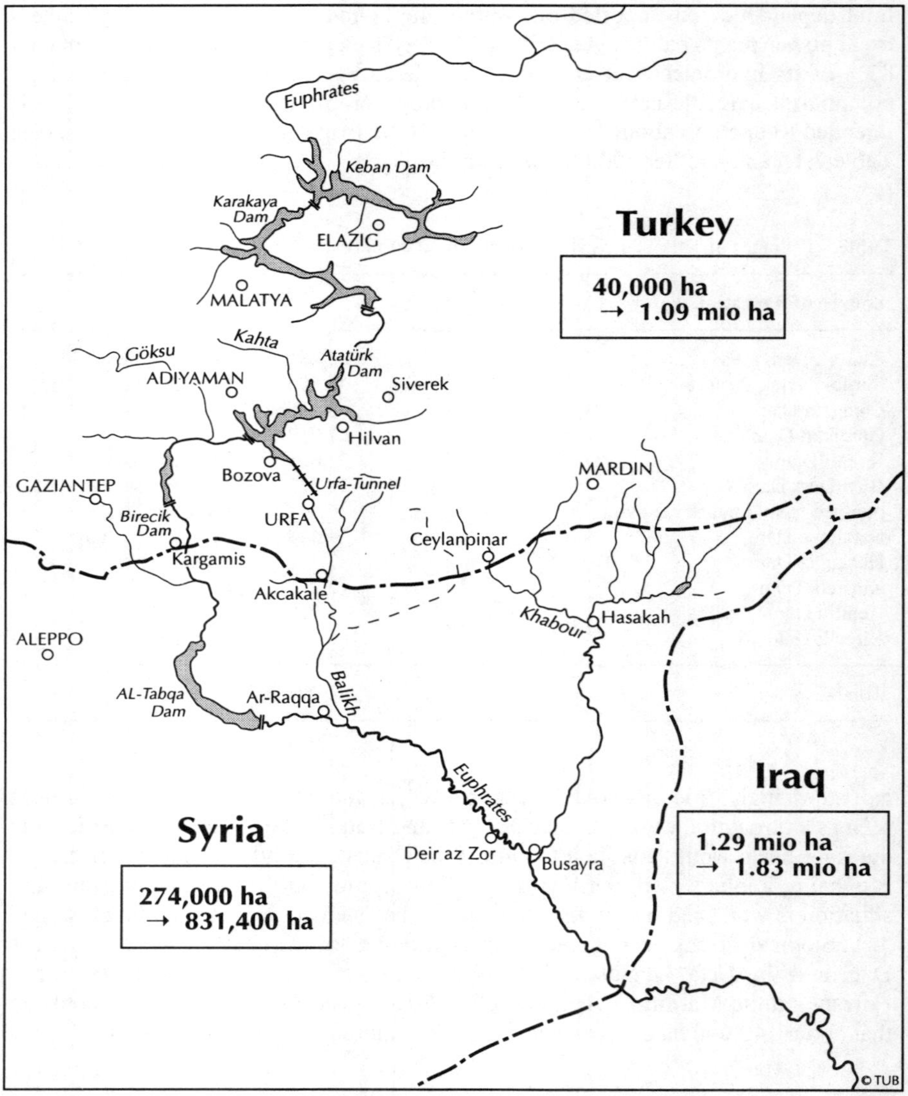

Figure 7.2. Planned and presently irrigated land in the Euphrates Basin

Prior to the completion of the Atatürk Dam, the Euphrates was mainly used in Iraq for irrigation, and to a smaller extent in Syria. In Iraq it is estimated that 1 to 1.29 million ha are irrigated, and that irrigated agriculture can be extended to 1.83 million ha (Al-Haditha Project). Syria started its development efforts with the construction of the Tabqa Dam in 1974; a second dam downstream, i.e. the Al-Baath Dam, was completed in 1986, and three more are planned on the Euphrates and its tributaries, i.e. the Al-Khabour and Balikh rivers. The Syrian Upper Euphrates Basin Project plans to irrigate 640,000 ha of additional land by the year 2000.[5] At the beginning of the 1990's, about 82,600 ha were equipped with the necessary infrastructure. It is estimated that in present day Syria a total of about 274,000 ha are irrigated by withdrawing water from the Euphrates and the Al-Khabour rivers.[6]

7.2 A Hypothetical Future Water Deficit

The envisaged realization of all irrigation projects bears a high potential for conflict because the amount of water needed will exceed the Euphrates' average annual discharge which is, measured at Hit/Iraq, 31.8 billion cubic metres (BCM).[7] All irrigation projects would use about 43.3 BCM,[8] and 7.4 BCM would evaporate from the reservoirs. The deficit would be 18.9 BCM which is, of course, an impossible scenario.

The Master Plan for the GAP estimated that the 1.091 million ha to be irrigated from the Euphrates would consume about 10 BCM and, therefore, would reduce the flow by one third. If evaporation from all Turkish reservoirs is considered, it is said that Syria would be able to realize only 42% of its projects, and Iraq 40%.[9] Syria's future water needs for the projects to be irrigated from the Euphrates and the Al-Khabour rivers are estimated to be 13 BCM, not including the yearly evaporation from the Tabqa reservoir, which is 1.5 BCM.[10] However, the withdrawal of water for the Turkish and the Syrian projects would reduce the water available for Iraq from 13-15 BCM[11] to approximately 7 BCM.

It is in any case difficult to anticipate future water demands for irrigated agriculture and other uses, but all calculations, although they use no commonly agreed figures, agree on the pessimistic scenario of negative effects for Syria's development plans and, in particular, for those of Iraq. Serious conflicts over the shared river seem to be inevitable because all countries, including Iraq, are pursuing policies of food self-sufficiency by extending irrigated lands. In 1990, Iraq imported 80% of its food from abroad, but it may „press for agricultural expansion as

[5] Chalabi and Majzoub 1995, p. 205. Wakil (1993) gives a figure of 799,000 ha (p.20).
[6] See Wakil 1993, p.19; Bakour 1992, p.154.
[7] The figure 31.8 BCM is used by Ubell (1971). Ünver (1997) gives the Euphrates flow at the Turkish-Syrian border as 31.8 BCM and Kolars (1992) 32.7 BCM.
[8] Return flows are not considered.
[9] Daoud 1996, p.5; see also Kolars 1992.
[10] Wakil 1993, pp.20-21.
[11] Kliot 1994, p.144.

a result of the economic boycott that the United Nations issued against Iraq following the Gulf War".[12] Turkey's and Syria's dependence on agriculture as a source of livelihood is high, providing 18% and 28%, respectively to gross domestic product; the agricultural sector employs 45% of the labour forces in Turkey, and 25% in Syria.[13]

7.3 The Present Shortage of Funds

The co-riparians' national plans for the extension of irrigated agriculture are, however, intentions, and not yet reality, and there are severe constraints in all countries on the implementation of irrigation schemes.

Although Turkish governments have, so far, heavily invested in the GAP region, Turkey no longer has the financial capacity to install the irrigation infrastructures in the envisaged time period.[14] There was already a delay in the timely completion of the Urfa Tunnel. A shortage of national funds caused the construction firm to not be paid on time and, therefore, the construction fell behind schedule. Irrigation priority schemes were selected to be completed in the year 2005, comprising 476,000 ha instead of 1.091 million. In the Urfa-Harran Plain, i.e. the first scheme to be irrigated, main and secondary delivery channels were already dug out, lined with concrete and ready for use at the end of the 1980's. As they remained unused for a couple of years, they now have to be renewed. In 1996/97, out of 146,000 ha, only 40,000 ha have been irrigated in a pilot area in this plain. Meanwhile, many farmers have continued to irrigate with ground water, dropping the ground water table of the aquifer which underlies the Turkish and Syrian territories. Recently, due to the lack of national funds, farmers have taken initiatives which might accelerate the path of infrastructure development. The farmers in the Bozova Plain have ordered the construction of irrigation networks from the state irrigation agency, which is financed with cheap credits provided by the Agricultural Bank to the farmers.[15]

The Syrian Upper Euphrates Development Project faces similar difficulties, i.e. public financial constraints. The project was started in 1974 with the aim to eventually irrigate 640,000 ha, but today only 82,600 ha are irrigable. The problems of salinization and gypsiferous soils appear to have seriously hampered the progress of soil reclamation; slow progress was mainly due to faulty planning and administration. In addition, studies by Syrian experts resulted in a uniform tendency to scale down the overall target. Present estimates of the area suitable for irrigation

---

[12] Kliot 1994, p.155.

[13] See Kliot 1994, pp.151-153.

[14] The GAP programme has reached a global financial realization rate of 40%. Out of the total public investment requirements of $32 billion, $12.5 has been invested from domestic resources; a total of $2.1 billion equivalent of foreign money has been allocated, and $1.5 have been credited for the Birecik Dam Project by a consortium of European firms. (Ünver 1997, p.197).

[15] See Milliyet, 28 June 1997, p.16.

range at 300,000 ha.[16] However, heavy investments are needed, and it is difficult to see where Syria will be able to find the necessary funds.

In Iraq, the greatest threat to expanding agricultural production is soil salinity: land has had to be abandoned and yields have declined by 30 to 50%. It is assumed that at present 80% of the irrigated land is affected by salinity. Rehabilitation efforts will need large amounts of water to get rid of the accumulated salt.[17] Investments for the reclamation of older schemes and for the construction of new projects are necessary, but it is unlikely that these investments can be undertaken unless the embargo is lifted, and even then, returns from oil exports have to be invested in other sectors.

In sum, the demands for the presently overall irrigated 1.6 million ha, together with the water requirements for the short- and mid-term plans of all co-riparian states (Figure 7.2.), do not exceed water availability because these demands are considerably below the average annual discharge of the Euphrates. Only if all, or most, planned irrigation projects are in full operation, will the likely demand exceed the water available. However, it appears that the national treasuries of the three co-riparian states are under higher pressure than the water resources.

Moreover, the national policies for the agricultural sectors heavily influence whether, in the long run, an acute water shortage will occur or not. At the national level there are many opportunities and options to reduce the potential for conflicts. As previously mentioned, plans can be changed with the result that the nations' demands decrease with the reduction of the areas to be potentially irrigated. In addition, and more importantly, all countries have high potentials for water savings within their agricultural sectors. There, water irrigation patterns cause great water losses of up to 40% due to methods of irrigation, high evaporation rates, seepage, and due to the fact that water is provided at zero cost.

Together with financial constraints, the rationing of water in irrigated agriculture by whatever means can ease the pressure on the water resource, but not necessarily the tensions. It seems to be that the governments maintain support from their political clientele by subsidizing irrigation water; if irrigation water is no longer subsidized and charged at its real, and not at zero costs, a political crisis on a national level could emerge, which the governments of all riparian countries would try to avoid by transferring the conflict to the international level. However, as water demands for irrigation gradually increase, a consensually agreed trilateral agreement is needed to provide a maximum of security for the nations involved.

7.4 The Negotiation Process: Water and Non-water Issues

The negotiation process over the depelopment of the Euphrates' water started between Turkey and Iraq in the 1940's. It was not before the 1970's when the patterns of conflict with which we are familiar today were born. The negotiated issues have changed over time, the coalitions have varied according to interna-

[16] See Lembke et al. 1990; Kliot 1994, p.139.
[17] Kliot 1994, p.159.

tional political circumstances, and outstanding factors have influenced water policies and were influenced by the water issue.

Concerning the Euphrates, it has become common practice to focus on conflicts more than on cooperative trends, giving the impression that tensions are overwhelming. It has not been properly addressed that cooperation could be forthcoming if the riparian states expected and gained benefits from an agreement, and if expected costs from a non-agreement were assumed to be too high. These benefits, or the costs, did not necessarily derive from water allocations but from other negotiated topics. Thus, cooperation was forthcoming not only in 'friendly' situations but also in situations where tensions were critical.

A brief analysis shows that cooperative trends dominated until the 1970's because no far-reaching projects were undertaken. The first agreement between Turkey and Iraq over the Euphrates was signed after World War II. In the Treaty of Friendship and Neighbourly Relations (1946),[18] both states agreed upon flood control measures to be undertaken in Turkey for the sake of Iraq. The parties to the treaty agreed that if the most suitable sites lie in Turkish territory, the entire costs will be financed by Iraq. Permanent observation stations will be built, operated and maintained by Turkey, in which Turkey and Iraq share in the costs in equal parts. Turkey agreed to inform Iraq of its plans for the construction works and, if Turkey will use water for purposes of irrigation and hydropower, this would be subject to separate negotiations.

The first controversy between Turkey and Syria was not over the Euphrates but over the Orontes River. When Syria, the upstream country, started with the construction of a dam in 1956 to drain a swamp and provide water for irrigation, Turkey reacted critically, as it was afraid that the Syrian dam would reduce the water available for the Turkish farmers. Some years later, Turkey built a dam of its own at the Orontes River.[19] Although there is an agreement, signed in 1939, which concludes that the water should be shared equally, Turkey blames Syria for using too much water, leaving zero water for Turkish farmers. The Orontes issue remains disputed, and Turkey wants it to be part of the Euphrates' negotiations, which Syria refuses.[20]

In the 1960's, initiatives for common projects alongside the Euphrates failed. Proposals for an integrated development of the whole Euphrates basin were impeded by the conflicting interests of the Eastern and Western Blocs, which supported one-sidedly either Syria's or Turkey's efforts.[21] Therefore, the nations started unilaterally developing and implementing dam projects. During the period it took to fill the Keban Dam (1974/75), i.e. the first dam on the Euphrates on Turkish territory, Turkey guaranteed a minimum release of 350 m^3/sec.[22] When Syria started with the flooding of the Tabqa Dam reservoir in the same year, Iraq wanted to fill the Habaniya lake for purposes of irrigation. When Iraq threatened to bomb the Tabqa Dam, the conflict escalated and brought the rival Ba'ath regimes

[18] United Nations Treaty Series 1949, 1. Nos. 574-582, Vol.37, pp.281-291.
[19] See Schulz 1995, p.103.
[20] Chalabi and Majzoub 1995, p.228.
[21] See Chalabi and Majzoub 1955, pp.198-201.
[22] This guarantee dates back to the year 1966.

of Syria and Iraq to the brink of war, adding to their competition for hegemony in the Arab world. According to the Iraqi government, millions of Iraqi farmers would suffer because of the reduced water flow. This severe conflict was reduced by the mediation of Saudi Arabia and the former Soviet Union,[23] and when Turkey increased the water release up to 450 m^3/sec.

The 1980's showed positive movements towards cooperation, and agreements were achieved. In 1980 Turkey and Iraq established the Joint Technical Committee for Regional Waters. The first meeting in 1982 was between Turkey and Iraq; Syria joined it in 1983. The three riparians agreed on the exchange of hydrological data. In the first meeting Turkey and Iraq declared to cooperate in the field of controlling the pollution of common waters and, more importantly, that the three governments should meet at a ministerial level „in order to decide on the methods and procedures to be recommended by the Joint Technical Committee to lead to a definition of the reasonable and appropriate amount of water that each country needs from the joint rivers".[24] Meanwhile, in 1983, tensions arose between Syria and Turkey when the water level in Lake Assad dropped and negatively affected Syria's electricity production. The Syrian government blamed Turkey for this shortage. Then, in summer 1986, Syria raised again its voice during the filling of the Turkish Karakaya Dam[25] to which Prime Minister Özal responded by ordering higher water releases for the downstream countries.[26] The filling period was not a conflicting issue between Iraq and Turkey because they had signed a protocol that Iraq would be informed of the filling programme, and it was guaranteed that this programme would take into account Iraqi needs from the Euphrates River.[27]

In 1987, a Protocol on Economic Cooperation was signed between Turkey and Syria.[28] It was the first time since 1923 that Turkish politicians had visited Syria; up to this time disputes over the province of Hatay, Turkey's NATO membership and its diplomatic relations with Israel impeded a rapprochement. The Protocol was the first bilateral agreement which dealt with water sharing, and the strongest formal agreement reached on regional waters since World War II. It guaranteed a yearly average release of more than 500 m^3/sec., or about 16 BCM per year, at the Turkish-Syrian border (see Box 7.1.).

---

[23] See Naff and Matson (eds.) 1984.
[24] Agreed Minutes of the First Session of the Turkish-Iraqi Joint Committee for Economic and Technical Cooperation 1980, in: Resmi Gazete 1981, No.17325, p.13;17.
[25] See Schulz 1995, p.104, quoting from Maoz and Yaniv 1986, pp.95-97.
[26] Nokta, 31 December 1989, Yil 7, p.52.
[27] Resmi Gazete 1984, No.18503, p.11.
[28] Resmi Gazete 1987, No.19660.

Box 7.1. Protocol on Matters Pertaining to Economic Cooperation Between the Republic of Turkey and the Syrian Arab Republic[29]

The Republic of Turkey and the Syrian Arab Republic,
Recalling historic and cultural ties existing between the two countries,
Desirous to add new dimensions to already existing good neighbourly relations,
Bearing in mind numerous complementaries in the economies of their countries,
agreed to sign the present Protocol. (...)

Water[30]

Art. 6: During the filling up period of the Atatürk Dam reservoir and until the final allocation of the waters of the Euphrates among the three riparian countries, the Turkish Side undertakes to release a yearly average of more than 500 m^3/sec. at the Turkish-Syrian border and in cases where the monthly flow falls below the level of 500 m^3/sec., the Turkish Side agrees to make up the difference during the following month.

Art. 7: The two Sides shall work together with the Iraqi Side to allocate the waters of the rivers Euphrates and Tigris in the shortest possible time.

Art. 8: The two Sides agreed to expedite the work of the Joint Technical Committee on Regional Waters.

Art. 9: The two Parties agreed in principle to construct and operate jointly projects in the lands of both countries on the Euphrates and Tigris rivers for irrigation and power generation provided that the technical and economic feasibility studies of these projects are carried out in cooperation by the experts of the two countries.

Art. 10: The Turkish Side explained the details of the „Peace Pipe Line“ planned to carry a portion of the waters of the Seyhan and Ceyhan rivers in Turkey through Syria by two pipe-lines, one going to countries of the Gulf and the other to the Hashemite Kingdom of Jordan and Kingdom of Saudi Arabia to supply water for household purposes and limited irrigation in the region. The Syrian Side agreed in principle to the project and showed interest provided that the Turkish Side carries out its technical and economic feasibility study by an international consultancy firm. The Syrian Side undertakes to facilitate the feasibility studies pertaining to the Syrian portion of the project. In case of its positive conclusion, the Syrian Side will enter into negotiations for the final realization of the project.

The English version of the Protocol was elaborated and signed in Damascus, on 17 July 1987.

Turgut Özal Turkish Prime Minister	Dr. Abdul Raouf El-Kassem Syrian Prime Minister

This agreement could be concluded because security matters and the topic of water were negotiated simultaneously. Prime Minister Özal, who was at that time the decisive political actor, committed to 500 m^3/sec. at the Turkish-Syrian border, which was more than before, with the intention of achieving an arrangement with Syria on security matters. Turkey accused Syria of supporting the Kurdish Workers' Party, the PKK, and wanted Syria to expel the PKK from their bases and out

[29] United Nations Treaty Series 87/12171, 17/7/1987.

[30] The Protocol further entails articles on Petroleum and Gas; Electricity; Trade; Banking Cooperation, and on Transport and Telecommunication.

of Damascus, the headquarters of the PKK-leader Öcalan. It was believed by the Turkish government that the PKK attacks, which started in 1984, would come to an end if Syria stopped its support. For a while it seemed that Özal's expectations were met, but a dramatic upsurge in fighting between Turkish forces and Kurdish insurgents in 1988/89 brought renewed expressions of concern from Ankara about Syrian support for the PKK, i.e. two PKK training camps in the Syrian-controlled Bekaa Valley in eastern Lebanon.[31] However, the Turkish government had 'mixed views' regarding the use of the 'water card' against Syria: the Turkish military and Prime Minister Özal were of the opinion that decreasing the amount of water could force Damascus to understand the importance of having 'good friendly ties' with Ankara while, on the other hand, the Turkish Foreign Minister insisted instead that „separate issues should not be put in the same basket".[32] Syria, on the other side, had successfully used the PKK as a trump card in the water game. When the Deputy Prime Minister of Iraq, Taha Yasin Ramadan, insisted on Iraq's participation and objected to the Protocol, Özal assured that its temporary character would be followed by a 'real treaty'.[33]

During the 1989 drought an agreement was signed between Ankara and Damascus to supply Syria with electricity rather than water because Syria's immediate need for water arose from low levels in the Tabqa reservoir. The generators were operating below capacity, and, except in Damascus, 5 hour power cuts were being imposed daily.[34] Irrespective of this agreement, hostility increased in Syrian-Turkish relations due to the shooting down of a Turkish survey aircraft inside the Turkish province of Hatay in October 1989 by Syrian MiGs. Disputes over this province date back to the 1930's when Hatay became, according to the French mandate, part of the Turkish nation over which Syria still claims territorial rights.[35] With regard to the Euphrates, the first event after the signing of the Turkish-Syrian Protocol was that Syria and Iraq objected to the date and duration for filling the Atatürk Dam in 1990/91. Before the inundation started, Iraq declared that if Turkey did not reduce the time from 4 to 2 weeks, Iraq would stop oil imports. Iraq feared that four hydropower plants which provide 40% of Iraq's electricity demands, would have to stop operation. As a result, an agreement between the Turkish Oil Company TÜRPRAS and the Iraqi ministry issuing 1.2 million tons of oil was not signed.[36] During the time it took to fill the dam, Iraq and Syria accused Turkey of not having released the guaranteed minimum amount (500 m^3/sec.). Turkey, instead, declared that it had notified the date and duration time. In the previous period, more water would have been released to fulfill the commitments of Article 6, and the filling period would already have been shortened and predated.[37] In addition, Turkey offered to compensate Syria for whatever shortfalls in

---

31 Financial Times, 23 October 1989; Neue Züricher Zeitung, 17 January 1990.

32 Turkish Daily News, 4 September 1989, p.B3.

33 Due to its nature, the Protocol was not ratified by Turkish parliament.

34 Water Power&Dam Construction, September 1989, p.3.
60% of Syria's power generation is produced by the Tabqa generators.

35 Financial Times, 3 January 1990.

36 Neue Züricher Zeitung, 18 January 1990.

energy generation caused by reduced water flows during the filling period, but there was no response from Damascus.[38] However, despite harsh rhetorics, the Syrian Vice Minister of Foreign Affairs, Yusuf Sakkur, stated in 1990 that the filling of the Atatürk Dam reservoir will have no negative impacts on the nations' bilateral relations, and the Iraqi Minister for Petrol, Tschelebi, declared that Iraq is sure that the GAP will provide benefits for the neighbouring countries as well.[39]

In April 1990, Syria and Iraq agreed that the water released at the Turkish-Syrian border be proportionally divided between Syria (42%) and Iraq (58%). At the meeting of the Joint Technical Committee in May 1990, Iraq demanded from the Turkish government an increase of the Euphrates flow up to 700 m^3/sec. as the 'final allocation pattern'; this was repeated at a summit meeting of the foreign ministers. Turkey insisted on 500 m^3/sec. as an average flow, but agreed to release 700 m^3/sec. for critical periods.[40] In February 1991, during the Kuwait War, the Turkish Prime Minister Özal „was encouraged to shut off the river to punish the Iraqis“ but „he replied that the Turks would not use water as a weapon“.[41] In 1995, a Syrian delegation headed by the Governor of Aleppo visited the Atatürk Dam site.[42] In the same year, Syrian and Iraqi objections were made against the construction of the Turkish Birecik Dam.[43] The Syrian and Iraqi governments wanted the Arab League countries to stop financial aid for Turkish projects and to boycott European firms that finance the Birecik Dam. They urged companies working on Birecik to pull out, and Iraq's under-secretary at the irrigation ministry said that these companies „could be shut out of future projects in Iraq and Syria, and perhaps in other Arab countries“.[44]

In April 1996, the Arab League demanded from Turkey an equitable allocation of the Euphrates' water, and that international financing agencies should only provide funding dependent on a fair distribution of water allocation.[45] From spring 1996 to summer 1997, the fundamentalist Refah Party with its leader Erbakan decided on foreign policy which centred on intensified relations with the Islamic world.[46] Amongst other initiatives, the Turkish Minister for Internal Affairs and the Minister for Energy and Natural Resources, both being members of the fundamentalist party, visited Bagdad and Damascus in August. They emphasized that there is enough water for all co-riparians.

---

[37] Turkish Daily News, 18 January 1990; T.C., Ministry of Foreign Affairs, Department of Regional and Transboundary Waters 1995.
[38] Financial Times, 3 January 1990; 15 January 1990.
[39] NewSpot, 18 January 1990.
[40] US Army Corps of Engineers 1991.
[41] Kolars 1994, p.65.
[42] Turkish Daily News, March 1995.
[43] It is the last dam on Turkish territory located near the Turkish-Syrian border.
[44] Financial Times, 15 February 1996.
[45] Neue Züricher Zeitung, 25 April 1996.
[46] The Refah Party came to power by coalition with Tansu Ciller's party, and was the first to nominate the Prime Minister.

Table 7.2. Agreed and conflicting issues between Turkey, Syria and Iraq

Year	Agreed issues	Conflicting issues	Parties
1946	flood control structures		Turkey and Iraq
1956		Syrian dam on Orontes River	Turkey versus Syria
1964	minimum release of 350 m^3/sec. (filling period Keban Dam)		Turkey and Iraq
1974/75 1975/76		filling period Keban Dam; filling period Tabqa Dam	Iraq versus Syria mediated by SU and Arab countries
1983		water level dropped in Lake Assad	Syria versus Turkey
1986	minimum release of 500 m^3/sec. (filling period Karakaya Dam)		Turkey and Iraq
1987	minimum release of 500 m^3/sec. at Turkish-Syrian border (filling period Atatürk Dam and until trilateral agreement)		Turkey and Syria
1990	proportional shares for Syria (42%) and Iraq (58%) from Turkish release		Syria and Iraq
		700 m^3/sec. instead of 500 m^3/sec.	Syria/Iraq versus Turkey
1991		date and duration time of filling period Atatürk Dam; minimum release not ensured	Iraq/Syria versus Turkey
1995/96		Birecik Dam	Syria/Iraq, Arab League versus Turkey

Source: Own compilation

The Turkish press interpreted this diplomacy that the Turkish Government would no longer insist on 500 m^3/sec. but on the Orontes issue. This was the first response to the Syrian and Iraqi request made in June 1995 which demanded that two-thirds of the Euphrates' average annual discharge measured at the Turkish-Syrian border be released to the downstream countries.[47] The recently established government under Mesut Yilmaz will certainly not continue with Erbakan's pan-Arabic policy which, however, completely failed when Ghaddafi, vis-a-vis Erbakan, stated that Turkey should stop suppressing the Kurds and finally allocate the Euphrates' water in an equitable manner.

The agreement concluded in 1987 and the continuing disputes are much more the outcome of outstanding issues than the result of a 'water crisis'. These issues have persuaded and impeded cooperative policies, and the unsolved water issue has remained useful for other national policies, such as the political intentions to achieve, or maintain, hegemony within the region, and security issues, i.e. to stop the PKK and other separatist movements from operating from the territory of another co-riparian. However, with respect to agreed commitments, there has been a long-enduring halt since 1987.[48] One obvious reason might be that Özal's expectations were not met; Syria has continued to support the PKK that has continuously been operating in southeastern Turkey from Syrian territory. Syria might try to re-establish its weakened position in the region by gaining support from Arab and European countries. It has been increasing pressure on the Turkish government via the Arab League and European countries that have, so far, friendly relationships with Turkey. It takes advantage of the fact that Turkey is concerned with obtaining full membership in the European Union, has important trade relations to Arab countries, is trying to extend economic opportunities in the region with the GAP and seeks credits for GAP-projects on the international market.[49] Iraq, as the downstream country, has the weakest position due to its location and no direct sanctioning power, which adds to its already politically weak role. While still maintaining good relations with Turkey, Iraq has formed a water coalition with Syria, although diplomatic relations have been at standstill since 1980.

[47] Al-Hayyat, 17 June 1995. This would result in a uniform tendency to scale down the size of all irrigation projects: Turkey could realize 68% of its planned irrigation projects, Syria 60 and Iraq 65% (Daoud 1996, p.6).

[48] The Joint Technical Committee held its final meeting in 1992.

[49] See Frey 1993, p.65.

7.5 Principle Controversies and the Water Allocation Procedures

Presently, the negotiated topics are whether 500 m^3/sec. (Turkey)[50] or two-thirds of the annual average mean flow (Syria, Iraq) should be the guaranteed minimum release at the Turkish-Syrian border; whether the Euphrates and Tigris rivers should be regarded as one hydrological unit (Turkey) or not (Iraq), and whether the Orontes River should be part of the negotiation process (Turkey) or not (Syria). Moreover, the co-riparians' understandings differ with respect to whether the Euphrates is an 'international' (Syria) or a 'transboundary' river (Turkey); whether a co-riparian can claim 'property rights' (Syria) or 'user-rights' (Turkey), and whether the water is to be 'shared' (Syria, Iraq) or to be 'allocated' (Turkey). Finally, the procedure and the criteria are disputed of how water shares can be determined. However, these underlying principal controversies are more disputed, rather than the amount of water to be released.

These issues, except the procedure and criteria, are based on the co-riparians rival doctrines of national sovereignty with respect to the Euphrates.[51] It seems that Turkish politicians favour the doctrine of absolute territorial sovereignty when they express that the Euphrates belongs to Turkey. The Turkish Minister of State, Kamran Inan, and, more recently, President Demirel expressed that Turkey possess the same right to the Euphrates' water as the Arabs have to their oil.[52] Syria follows this doctrine with regard to the Orontes River. However, this doctrine is obsolete, and rejected by almost all river treaties, as one nation cannot use the water within its territory without restrictions. Iraq and, in some respects, Syria claim rights-of-prior use, which refers to the doctrine of absolute territorial integrity. This doctrine is also obsolete, as it implies that a downstream riparian, having developed the water resources first, could exercise a veto towards later development activities of an upstream country. The doctrine of limited territorial sovereignty is widely accepted, and almost all river treaties rely on them. Despite their controversies on sovereignty, all co-riparians refer to the principle of 'equitable use',[53] based on the Helsinki Rules on the Uses of the Waters of International Rivers (1966) of the International Law Association, and on the principle of good neighbourhood. The Helsinki Rules emphasize equitable utilization, i.e. each basin state is entitled to a reasonable and equitable share in the beneficial uses of shared water, and that actions should be prevented within one state's border that would significantly harm other states. It permits harmful actions but requires compensa-

[50] It is important to note that the Protocol is questioned by Turkish lawyers; because it has never been ratified by parliament, they regard it as a non-binding agreement (Welt-Rundschau, 29 April 1996).

[51] On the doctrines see Barandat and Kaplan in this volume.

[52] See Chalabi and Majzoub 1995, p.208; Barandat 1995, p.35.
Kamran Inan was close to the Turkish Prime Minister Turgut Özal who stated that water will not be used as a weapon against the Arab countries.

[53] See Solanes 1992, p.117.

tion or mitigation as acceptable alternatives to avoidance, and the obligation to notify and inform other nations of any activities on shared watercourses that will affect them.[54]

However, the definition and quantification of a 'reasonable' and 'equitable' share is disputed. In 1983, the co-riparian states agreed that the three governments should meet at a ministerial level in order to decide on the methods, procedures and criteria to lead to a definition of the reasonable and appropriate amount of water for each country. More than ten years later, the countries are far from a common denominator. Their proposals reflect the stages of water resources development and they support, not surprisingly, individual instead of common interest.

In 1984, Turkey proposed its „Three-stage plan for optimum, equitable, and reasonable utilization of the transboundary watercourses of the Tigris-Euphrates Basin", which comprises inventory studies for water resources (first stage) and land resources (second stage), and the evaluation of both water and land resources (third stage), which includes the discussion of methods and criteria for determining economic viability of the planned projects.[55] The inventory studies would not only provide the same sets of data for all parties involved, but make it possible to plan water use more efficiently. All opportunities for water savings for the realized and for the planned projects should be taken into account. This point is crucial, as it implies that „after the preparation of a standardized soil classification of lands that could be irrigated (...), it seems reasonable to plan to irrigate higher quality lands *irrespective* of the country in which they may be located".[56]

Syria and Iraq use mathematic formulae in order to define definite water quotas. Syria proposed that the co-riparians should declare their demands for each river separately, i.e. the Tigris and the Euphrates rivers. If the claims exceed a river's discharge, the deficit will be proportionally deduced from each share. The Iraqi mathematic formula is somehow different and admits that each riparian should declare its claims for the realized projects, then for those under construction and, finally, for the planned. The nations' water quota would be defined subsequently, i.e. first for the projects in operation, then for those under construction, etc., with the realized projects having priority over the rest in favour of Iraq's early development efforts.

The Turkish proposal with its basin-wide approach follows internationally accepted recommendations, and starts from a crucial pre-requisite, i.e. the collection and exchange of data at commonly agreed gauging stations, which has not yet been arranged. However, its centrepiece is not consensual: it can be assumed that the most suitable lands to be irrigated lie on Turkish territory, where higher efficiency gains could be achieved. The selection of these sites would force Syria and Iraq to partly give up their development plans, and, moreover, to accept a higher degree of dependency on Turkish food imports. The Syrian and Iraqi prosposals, instead, insist on the apportionment of the average annual flow, and one may argue that the riparians would then claim absolute sovereignty over their share, opposing inter-

---

[54] From Gleick 1993, pp.106-108.

[55] See T.C., Ministry of Foreign Affairs, Department of Regional and Transboundary Waters 1995, pp.35; Turan 1993, p.23.

[56] Turan 1993, pp.28-29 (emphasis added).

national rules.[57] The Syrian and Iraqi proposals are more pragmatic and do not afford such an intensive degree of cooperation and integration, but they one-sidedly favour and shield their national development plans.

However, if one considers the co-riparians' relationships, it seems that a basin-wide approach is not the best solution. In the case of the Euphrates, where there is great mistrust among the co-riparians, pragmatic approaches would consider a high degree of sovereignty.

7.6 Non-negotiated Issues

There are a number of issues which have so far not been addressed in the negotiation process. They refer to if and how to compensate for investments when upstream activities have positive impacts on downstream countries; how to share the risk of low water conditions, and water quality matters.

7.6.1 Positive Impacts of Upstream Activities on Downstream Riparians

It is the negative effect of reduced water flows over which conflicts continue among the riparian states. However, the construction of dams in an upstream country has not only negative effects on downstream riparians:[58] A dam regulates a river's discharge and reduces peak flood flows through temporary storage. This is particularly important for avoiding damage which results from the inundation or from the flood water itself. Rivers contain sediment which partly accumulates in the reservoirs. Depending on sediment loads and a reservoir's storage capacity, the accumulation reduces, over time, its safe yield, i.e. the annual runoff of a river which can be utilized for electricity generation, irrigation and flood control.[59] If an upstream reservoir serves as a sediment trap for those downstream, it increases the life time over which they provide services; it reduces maintenance requirements[60] in irrigation channels because less sediment deposits therein. In addition, water quality improves downstream with less sediment contents.

The Turkish Keban Dam has positive flood control effects for the downstream countries, in particular, for Iraq. Positive effects on flood control increased with the construction of the Karakaya and Atatürk dams, which also serve as sediment traps. The deforested Turkish watershed of the Euphrates has extremely negative effects on sedimentation rates, and estimates assume that yearly sedimentation in

[57] E.g. Kibaroglu (1996, p.34) emphasizes that the Turkish proposal would apply to the principle of limited territorial sovereignty while the Syrian and Iraqi do not.

[58] See Table 7.3.

[59] One example is the Turkish Seyhan Dam; its active capacity was reduced by 60 MCM between 1956 and 1986 with negative consequences for its flood control capacity. Experiences from other countries are reported in Brown and Wolf 1984 and Xuemin 1992.

[60] I.e. sediment removal.

the three Turkish reservoirs, i.e. Keban, Karakaya and Atatürk, can reach a volume of 1,050 cubic metres per square kilometre.[61] Erosion control measures are part of the GAP and afforestation has been undertaken on approximately 50,000 ha. However, the high storage capacities of the Turkish dams are of extreme importance for the Syrian Al-Tabqa reservoir which has, due to topography, a smaller storage capacity.[62] Therefore, the beneficiaries should share in investment cost, costs for resettlement, erosion control measures and operation and maintenance of the hydraulic infrastructures.

The Turkish Birecik Dam is, however, different. It serves irrigation purposes (80,000 ha), energy generation and the regulation of peak floods, which are produced when the Atatürk hydropower plant operates at peak demand. The 8 turbines of the Atatürk plant, each of them having a discharge of 225 m^3/sec., might produce floods of about 1,800 m^3/sec. These peak floods would cause appreciable harm to the detriment of Syria if they discharged without regulation. The Syrian Al-Baath Dam, downstream from Tabqa, serves the same purpose. Because these dams simply correct negative effects which derive from development efforts within the same country, they cannot be subject to compensation.

Table 7.3. Negative and positive impacts of upstream activities on downstream riparians

Negative impacts	**Positive impacts**
control over water flows reducing water flows peak floods (energy generation) water quality	reducing peak flood flows trapping sediment storage for low water conditions

7.6.2 Low Water Conditions

The high storage volumes of the Turkish reservoirs,[63] in particular that of the Atatürk Dam, allow for the regulation of water flows and make, to a certain degree, the release of steady flows in periods of low water conditions possible. In the Euphrates' watershed, severe droughts occurred from 1958 to 1962 and from 1970 to 1975, when the annual discharge fell to 49 and 62% of the annual mean respectively. Again, in 1989, the worst drought in 50 years was recorded.[64] Since 1937,

---

[61] The General Directorate for State Hydraulic Works assumes 350 cubic metres per square kilometre for each reservoir.

[62] The Tabqa reservoir has a maximum capacity of 11.6 BCM, the Atatürk Dam stores 48.7 BCM.

[63] All Turkish reservoirs at the Euphrates have an active capacity of approximately 42 BCM.

[64] The costs for Turkey were reported to account approximately US$ 2 billion. (Financial Times, 19 October 1989; 29 March 1990).

when the historical runoff on the Euphrates was documented, in more than 20 years the discharge has been less than the yearly average. Concerning this, the Turkish-Syrian Protocol (1987) does not provide adequate solutions because it guarantees a minimum release in any case, completely ignoring precipitation patterns such that the risk is spread unequally among the riparian states. According to the Protocol, Turkey would bear the costs in periods of long-enduring droughts, i.e. it would have to reduce its own water use, adding to the already high investment costs. The Syrian-Iraqi agreement (1990) committed to proportional shares and considered low water conditions, whether this was intended or not, and naturally caused risks are shared equally among the parties. If a future trilateral agreement committed to water quota, proportional shares would appear to be a much more balanced and equitable solution.

In addition, the minimum release of 500 m^3/sec., which is guaranteed by the Protocol, would not have been possible without the storage capacity of the Atatürk Dam. Syria benefits from the steady flow, and in this respect, the Turkish dams reduce uncertainty in water supply for the downstream countries.

7.6.3 Water Quality

Up to now, water quality problems have not been negotiated, despite the fact that the quality of water suitable for irrigation will decrease with agricultural usage. Because the nations' development plans centre on irrigated agriculture, they need at least water quality suitable for irrigation purposes. The GAP will certainly change the water quality because of its industrial, urban and agricultural development. Textile and leather industry, food, fodder and meat processing will develop accompanied by the expansion of urban centres. The impact on the water released downstream depends on whether treatment plants are operated or not, and cannot yet be assessed. However, even if industrial and communal waste water is treated and usable for irrigation, water quality will deteriorate due to agricultural effluents. So far, negative impacts are limited, at least those from the irrigated Turkish sites.[65] In the future, however, it will be difficult to measure the irrigation-induced salinity content of the irrigation water because, even without irrigation, salinity has been increasing due to decreasing precipitation and high rates of evaporation. In the upstream basin at the Keban gauging station, an average electrical conductivity, i.e. the indicator for salt concentrations, of 407 micromhos/cm was measured; it increased in Syria, and reached its maximum value in low-flow months of more than 1,600 micromhos/cm at Hit (Iraq).[66] These values date back to a time where irrigation was neglible in Turkey and Syria; they certainly will get worse. A problem of particular concern poses the discharge of drainage water when the Turkish schemes are irrigable: the salinity concentration of drainage water may exceed that of irrigation water four fold; there is no natural outlet and the disposal of the highly salt loaded drainage water needs an appropriate solution. Apart from in-

[65] Within the boundaries of the GAP, only 40,000 ha are opened up for irrigation.

[66] See Franke 1979; Government of Iraq Development Board, Ministry of Development 1957/58; Scheumann 1993.

creasing salt concentrations, pesticide usage poses a threat to water quality and may make it unsuitable for human consumption (drinking). However, Syria and Iraq will be affected differently by contaminated water: the victim will be Iraq because the river transports chemicals from the Turkish and Syrian projects.

7.7 Negotiating Positive Impacts

The Helsinki Rules of the International Law Association and the recently adopted International Water Convention agree that international rivers should be shared equitably and reasonably, and that water resources should be used in an optimal manner. This would include an equitable allocation of costs and benefits, and that positive impacts are subject to compensation which has not yet been considered in the convention. If the downstream countries are beneficiaries, they should share in the costs. One positive example is the Turkish-Iraqi Treaty (1946), where Iraq as the beneficiary committed to bearing the costs of flood control measures. An agreement between India and Nepal included a similar commitment: the downstream country financed the costs of an upstream dam which was built for its sake.[67] In the negotiation process over the rivers Usumacinte, Suchiate and the Grijalva between Guatemala and Mexico, Mexico was of the opinion that the upstream country Guatemala should be compensated for the costs of flood control measures by the downstream country. The Rio Grande treaty concluded in 1933 between Mexico and the United States, admits that the costs of irigation and flood control means are to be shared in proportion to the irrigated and protected lands. With respect to the Colorado River, Mexico and the U.S. arranged a cost-sharing agreement varying with the uses: the investment costs of flood control measures are evenly shared; the costs of other means relate to property rights on water, which in turn are defined on how much water is provided by one country's basin and stored in their reservoirs.[68] In the Columbia Treaty, the U.S. compensates the Canadian state for its costs of flood control infrastructures.[69]

These treaties demonstrate that compensation, or costs, are negotiable only if particular conditions are given:

- Infrastructure projects are commonly undertaken, and costs are arranged beforehand, and
- the parties to a treaty have common benefits, such as the control of floods.

If projects are undertaken unilaterally, and if positive effects are more or less a by-product, e.g. the regulation of a river's discharge and the storage of water for periods of low water, claims for compensation are difficult if not impossible. Moreover, it is unlikely that a negotiating party would commit to compensation for positive impacts unless there is an agreement on negative effects. However, consider-

[67] See Lichem 1982.
[68] See United Nations 1987, pp.81.
[69] See United Nations 1987, p.89.

ing positive effects cannot simply compensate for negative impacts as they refer to different uses.

In the case of the Euphrates, positive effects from upstream activities have, so far, not been part of the negotiations; on the contrary, if Turkey demanded compensation, the situation would certainly heat up. Presently, common water development projects have little chance, although the Turkish-Syrian Protocol intended that „the two parties agreed in principle to construct and operate jointly projects in the lands of both countries for irrigation and power generation".[70] Furthermore, the Syrian government agreed to the Peace Pipeline, which is to deliver water from the Turkish rivers Seyhan and Ceyhan to Middle East countries;[71] later on, Syria refrained because the project would include water supply for the Israelis.

Presently, there is no solution to the disputed Euphrates which is acceptable to all parties involved, but other factors such as trade relations may support cooperative trends between the riparians: Turkey is Syria's fifth largest trading partner, and total trade between Turkey and Syria more than doubled between 1988 and 1992. Prior to the embargo, Iraq and Turkey had well developed trade relations, and Turkey was Iraq's major supplier of goods, both from domestic production and from transit trade.[72] Moreover, both Arab states signed treaties with Turkey to overcome deficits in energy generation; a maximum of 400 million kWh can be exchanged annually among the partners.[73]

7.8 Conclusions

At present, there is no acute water shortage, and even the realization of the projected mid-term plans poses no threat to the amount of water available. In addition, there is a high potential for water savings within the agricultural sectors of all co-riparian states which can, in the long run, ease the pressure on the resource, although not necessarily the tensions. Presently, security issues are crucial between Turkey and Syria, and tensions have so far derived from national security perceptions and political intentions to maintain hegemony within the region. This adds to the disputed water issue and creates the particularly high potential for conflict between the riparian states. However, an agreement is possible, irrespective of the states' imperative on sovereignty if the parties involved benefit from an agreement. Historical experiences suggest that benefits do not necessarily have to be generated from water allocations; non-water issues promoted the 1987 Turkish-Syrian arrangement, when decisive Turkish politicians expected high costs from a non-agreement. If compared to the above mentioned water treaties, there are more than two parties, and the strategic behaviour of one party depends not only on national but on regional politics. However, Syria and Iraq already have agreed on proportional shares, and their requested water quota and the Turkish offer are close to

[70] Resmi Gazete 1987, No.19660.

[71] See Box 7.1. for Art.10.

[72] Financial Times, 14 June 1988; Nachrichten für Aussenhandel, 31 May 1990.

[73] Nachrichten für Aussenhandel, 28 June 1988.

agreement. Although definite water quotas are not the best solution, sharing might be a viable solution for the Euphrates' riparian states because it limits restrictions on sovereignty. Even then, cooperation for regulations on low water conditions, eventually deterioriating water quality, measurement of data, control and operation of flows remains a necessity.

References

Bakour, Yahia 1992: Planning and Managing of the Water Resources in Syria, in: LeMoigne et al. (eds.), Country Experiences in Water Resources Management. World Bank Technical Paper No.175, Washington, D.C.

Barandat, Jörg 1995: Wasser - regionaler Konfliktstoff weltweiter Bedeutung. Hamburger Beiträge zur Friedensforschung und Sicherheitspolitik, Heft 96. Hamburg.

Brown, Lester; Wolf, Eric1984: Soil Erosion - Quiet Crisis in the World Economy. Worldwatch Institute Paper No.60. Washington, D.C.

Chalabi, Hasan; Majzoub, Tarek 1995: Turkey, the Water of the Euphrates and Public International Law, in: Allan and Mallat (eds.), Water in the Middle East. Legal, Political and Commercial Implications, pp.189-233. Tauris Academic Studies, I.B. Tauris Publishers London and New York.

Daoud, Maged 1996: Regional Water Resources and Their Influence on the Economy of Syria. Stiftung Wissenschaft und Politik, Forschungsinstitut für internationale Politik und Sicherheit.

Franke, Christian 1979: Limnological Investigations on Lake Assad. Al Thawra.

Frey, Frederick W. 1993: The Political Context of Conflict and Cooperation Over International River Basins, in: Water International, 18, pp.54-68.

Gleick, Peter 1993: Water and Conflict. Fresh Water Resources and International Security, in: International Security 18 (1), pp.79-112.

Government of Iraq Development Board, Ministry of Development 1957/58: Study of the Lower Tigris-Euphrates Rivers. Baghdad.

Kibaroglu, A. 1996: Prospects for Cooperation in the Euphrates-Tigris Basin. Cooperation Within International River Basins, in: Howsam, P.; Carter, R.C. (eds.), Water Policy: Allocation and Mangement in Practice, Proceedings of International Conferene on Water Policy, held at Cranfield University, 23-24 September 1996, pp.31-38. E&FN SPON, London, Weinheim, New York.

Kliot, Nuri 1994: Water Resources and Conflicts in the Middle East. Routledge London and New York.

Kolars, John F. 1992: The Future of the Euphrates Basin. World Bank Technical Paper No.175, Washington, D.C.

Kolars, John F. 1994: Problems of International River Management: The Case of the Euphrates, in: Biswas (ed.), International Waters of the Middle East. From Euphrates-Tigris to Nile, pp.44-94. Oxford University Press, Bombay, Delhi, Calcutta, Madras.

Kolars, John F.; Mitchell, William A. 1991: The Euphrates River and the Southeastern Anatolia Development Project. Southern Illinois Press, Carbondale and Ewardsville.

Lembke et al. 1990: The Euphrates Development Scheme in Syria. Social Impact, Production and Linkages, Berlin.

Lichem, Walter 1982: Institutional and Legal Aspects Related to National Multi-Purpose Water Resources Projects on International Rivers or Interstate Rivers, in: Water for Human Consumption, Vol.1., Dublin.

Naff, Thomas; Matson, Ruth (eds.) 1984: Water in the Middle East. Conflict or Cooperation? Westview Press, Boulder and London.

Scheumann, Waltina 1993: New Irrigation Schemes in Southeast Anatolia and in Northern Syria: More Competition and Conflict over the Euphrates?, in: Quarterly Journal of International Agriculture, Vol. 32., No.3.

Schulz, Michael 1995: Turkey, Syria and Iraq: A Hydropolitical Security Complex, in: Ohlsson, Leif (ed.), Hydropolitics: Conflict over Water as a Development Constraint. London.

Solanes, Miguel 1992: Legal and Institutional Aspects of River Basin Development, in: Water International, 17, pp.116-123.

T.C., Ministry of Foreign Affairs, Department of Regional and Transboundary Waters 1995: Water Issues between Turkey, Syria and Iraq. Ankara.

Turan, Ilter 1993: Turkey and the Middle East: Problems and Solutions, in: Water International, 18, pp.23-29.

Ubell, K. 1971: Iraq's Water Resources, in: Nature and Resources 7, 2.

United Nations 1987: Institutional Issues in the Management of International River Basins: Financial and Contractual Considerations. Natural Resources/Water Series No. 17, New York.

United Nations Treaty Series 1949, 1.Nos. 574-582, Vol.37.

Ünver, Olcay 1997: Southeastern Anatolia Integrated Development Project (GAP), Turkey: An Overview of Issues of Sustainability, in: Water Resources Development, Vol.13, No.2, pp.187-207.

Wakil, Mikhail 1993: Analysis of Future Water Needs for Different Sectors in Syria, in: Water International, 19, pp.18-22.

Xuemin, Cheng 1992: Reservoir Sedimentation at Chinese Hydro Schemes, in: Water Power&Dam Construction, October.

Newspapers

Al-Hayyat, 17 June 1995.

Financial Times, 14 June 1988; 19 October 1989; 23 October 1989; 3 January 1990; 29 March 1990; 15 February 1996.

Milliyet, 28 June 1997.

Nachrichten für Aussenhandel, 28 June 1988; 31 May 1990.

Neue Züricher Zeitung, 17 January 1990; 18 January 1990; 25 April 1996.

NewSpot, 18 January 1990.

Nokta, 31 December 1989, Yil 7.

Resmi Gazete 1981, No. 17325.

Resmi Gazete 1984, No.18503.

Resmi Gazete 1987, No.19660.

Turkish Daily News, 4 September 1989;18 January 1990; March 1995.

Water Power&Dam Construction, September 1989

8 Conflicts over the Nile or Conflicts on the Nile?

Manuel Schiffler

„In playing with water Sudan is playing with fire", declared Egyptian Foreign Minister Amre Muhammad Mussa when, in August 1995, Sudan threatened to withdraw from the 1959 treaty between the two states on the allocation of the waters of the Nile. The Egyptian government had previously accused Sudan's Islamist government of responsibility for an attack in the Ethiopian capital of Addis Ababa, which nearly claimed the life of Egypt's President Hosni Mubarak. Sudan's threat to use the 'water weapon' was a response to the ensuing Egyptian recriminations.

Plans on the part of any of the nine states on the upper course of the Nile to more intensively harness the Nile waters have always evoked nervous reactions among Egyptian governments. Apart from some nonrenewable ground water reserves in remote desert regions, Egypt has no water resources of its own, and it is thus nearly wholly dependent on the Nile for its water supplies. The ancient Greek Historian Herodotus had already aptly characterized Egypt as a 'gift of the Nile'. Winston Churchill described Egypt as a country that depends on the Nile the way a diver does on his air hose. Accordingly, Egypt's concern about guaranteed flows of Nile water is a constant factor of its foreign policy. During his tenure of office as a Minister of State in the Egyptian foreign ministry, Butros Butros Ghali went so far as to warn repeatedly that the next war in the Middle East would be fought not for oil but for water.[1]

Besides Egypt, the main actors in the conflict over the waters of the Nile are Sudan and Ethiopia. Some 85% of the Nile waters rise in Ethiopia and flow, via the Blue Nile and the Atbara, through Sudan into Egypt. Only 15% of the Nile waters come from the White Nile, which is fed from rainfall around Lake Victoria. Aside from Sudan, the states bordering on the White Nile include Zaire, Burundi, Rwanda, Tanzania, Kenya, and Uganda (see Figure 8.1.). In addition, parts of the Atbara, a tributary of the Nile, rise in Eritrea. Compared to the three main actors, however, these states play no more than a subordinate role in the conflict over the waters of the Nile.[2]

---

1 Warburg 1991, p.572.

2 See Kabanda and Kahingere (1994) for proposed irrigation projects in Uganda using 6 billion cubic metres (BCM) per year for irrigation by 2050.

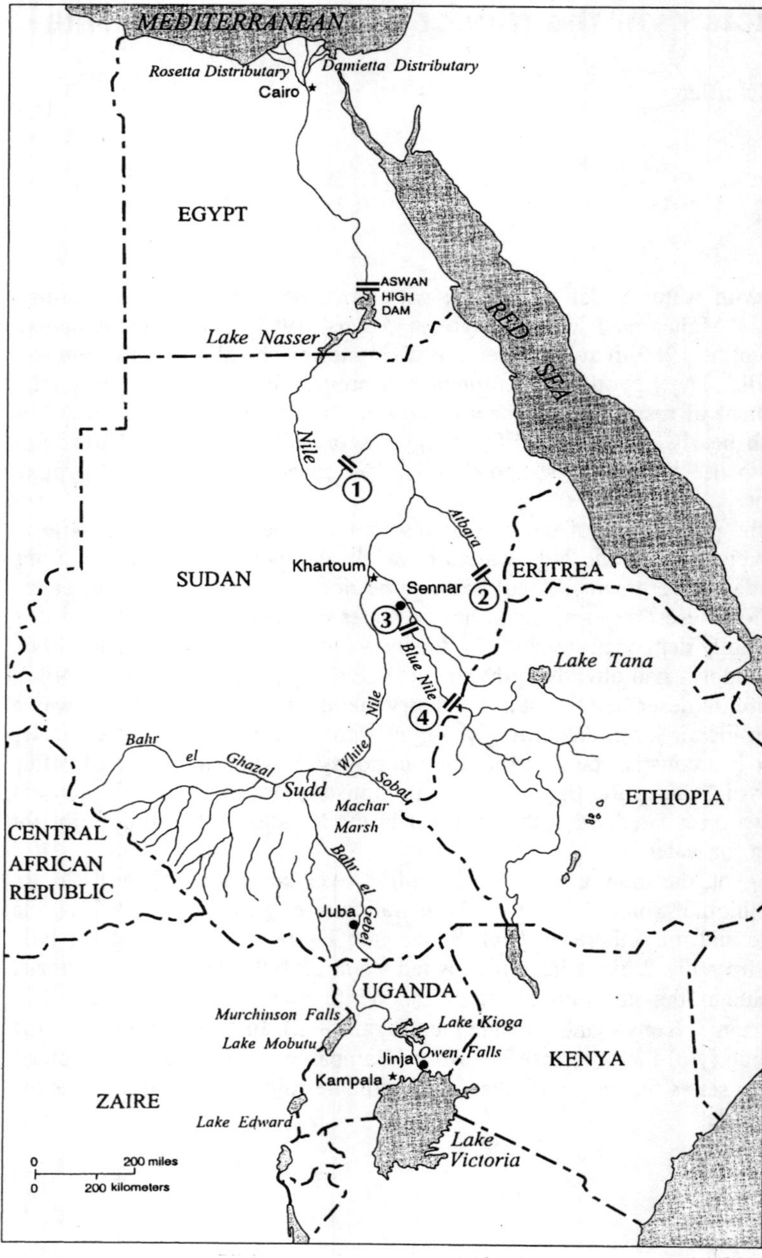

Figure 8.1. The Nile Basin

8.1
The Century Water Storage Scheme

The history of the attempts to manage the waters of the Nile within the river's entire catchment extend back to the era of the British colonial empire. It was in 1904 that British engineers first proposed the Century Water Storage Scheme, which was to be further elaborated in subsequent years, and whose aim was to collect the floodwaters of the Nile outside Egypt's borders. This was intended to ensure a constant annual flow. Droughts and floods were to be relegated to the past, and the water available in Egypt for irrigation purposes was to be increased substantially.[3] The scheme provided, among other things, for the construction of a dam on Lake Tana in Ethiopia at the sources of the Blue Nile, a dam on Lake Victoria at the sources of the White Nile, a series of dams in Sudan, and a canal through the swamps of the Sudd in southern Sudan; these measures were intended to increase the quantity of water available by lowering evaporation losses. Due to the huge evaporation losses anticipated, which would be much lower in the Ethiopian highlands, there were no plans to build any dams in Egypt. The collapse of the British colonial empire, and a lack of funds at that time prevented any implementation of the overall scheme. Yet at least individual components of the scheme were later either realized (e.g. construction of the Owen Dam on Lake Victoria in the 1950's) or started on (e.g. the Jonglei Canal in southern Sudan at the beginning of the 1980's).

8.2
First Agreements on the Waters of the Nile

As early as 1902 Ethiopia had undertaken, in an agreement signed by Emperor Menelik II, to erect dams on the Blue Nile only in mutual agreement with Great Britain and Sudan.[4] This agreement, whose significance is occasionally pointed to by Egypt today, was never ratified by Ethiopia and is hence regarded by the latter as invalid.[5]

The Egyptian nationalist Sa'ad Zaghlul spelled out the Nile's significance for Egypt by contending that Sudan is more important to Egypt than Alexandria. The British high commissioner of Egypt and Sudan, Lord Edmund Allenby, fanned the flames of Egyptian anxieties with regard to the Nile by threatening, following the assassination of the governor-general of Sudan by an Egyptian nationalist in 1924, to cede to Sudan as much water as it wished, without taking Egypt's rights into account.[6] At this juncture the Senar Dam on the Blue Nile was under construction; it serves to irrigate the cotton fields in the Gezira to the south of Khartoum. In Egypt, threats like those pronounced by Lord Allenby have contributed to the

---

[3] Waterbury 1979, pp.61; Kliot 1994, pp.35-37.
[4] Waterbury 1990, p.280; Okidi 1994, pp.323-325.
[5] Ayeb 1992, p.68.
[6] Warburg 1991, pp.568.

traumatic fear that any master of Sudan could use the waters of the Nile to blackmail Egypt. In fact, however, the course taken by the history of Nile water allocation in the subsequent years did not support that fear.

The year 1929 experienced an exchange of diplomatic notes between a now-independent Egypt and the British high commissioner Lord Lloyd, acting in the name of Sudan, then under joint British and Egyptian administration. The exchange of notes concerned the allocation of the Nile waters between the two countries: Egypt was to receive 48 billion cubic metres (BCM) of water per year, while Sudan was accorded only 4 BCM.

8.3 The Nile Water Agreement of 1959

When Sudan was set to gain its independence, the so-called Unionists at first agitated in Sudan for a union with Egypt. But they lost out to their political opponents under Ismail al-Azhari, who led Sudan to independence in 1956.[7] The government of Sudan rejected the 1929 'treaty', demanding an increase in the share of water to which Sudan was entitled. At the same time, Gamal Abdel Nasser, who was forging ahead with the construction of the Aswan High Dam, was interested in an internationally binding and secure guarantee for Egypt's share of the waters of the Nile. Part of Lake Nasser, as it was to be named later, would inundate Sudanese territory and Sudan claimed compensation for the resettlement of its citizens. Tenacious negotiations led on November 8, 1959, to the signing of the Agreement for the Full Utilization of the Nile Waters between Egypt and Sudan; it has until today been observed and is in many aspects regarded as a model for the allocation of water by mutual consent on international rivers.[8] The agreement stipulates that Egypt is to receive 55.5 BCM of water, while Sudan has a right to 18.5 BCM.[9] It was assumed that in addition 10 BCM would evaporate in Lake Nasser (this in itself amounts to more than three times the drinking water consumed by Egypt today!); in other words, the entire flow of the Nile, on average 84 BCM, was to be divided up at Aswan.

The water allocated to Egypt has until today been sufficient to cover Egypt's consumption. The figures about Egyptian water use cited in different sources considerably differ, but there is agreement that most of the Nile water is consumptively used in Egypt. Less than one tenth of the Nile water entering Egypt today still flows into the Mediterranean. On the average, this flow currently amounts to some 5 BCM per year flowing past the Edfina barrage on the Rosetta branch of the Nile into the Mediterranean during the winter months.[10] The remaining water either evaporates in the canals and on the fields or evapotranspirates through plants. Irrigated agriculture accounts for far more than 90% of the water consumption in

[7] Warburg 1991, p.570.

[8] For some weaknesses in the agreement, see the first contribution by Schiffler in this volume.

[9] United Nations 1963; Okidi 1994, pp.333-335.

[10] Stoner 1994, p.202.

Egypt. As the actual average inflow of water into Egypt was higher than the share allocated under the 1959 agreement, total current consumptive water use in Egypt probably is close to its total share.

The 1959 Agreement also provided for the formation of a Permanent Joint Technical Committee, a body made up equally by Egyptians and Sudanese and having an alternating chairmanship. The Committee, which is headquartered in Khartoum, monitors compliance with the accord. Despite all political tensions between the two countries, the Committee has continued with its work during the past 37 years.[11] The agreement furthermore provides for construction measures aimed at increasing the flow of the Nile; their costs are to be shared equitably by the two signatory states. The additional water obtained in this way is likewise to be allocated equitably. What this provision had in mind was above all the project of building a canal through the Sudd in southern Sudan. Talks with other Nile states on their water rights are to be conducted jointly by the two signatories. Any diminution of water rights resulting from such talks are to be borne equally by both signatories.[12]

Following conclusion of the accord, the Aswan High Dam was built in Egypt between 1960-70. It was the first dam on the Nile that was large enough to store all of the Nile's floodwaters. Roughly at the same time the Roseires Dam, smaller in comparison, was built on the Blue Nile. It was, however, at that juncture already evident that Sudan would be unable to expand its dams and irrigation network quickly enough to be able to profit fully from its water rights. The surplus water unused by Sudan automatically benefits Egypt. The figures of the amount of water actually used by Sudan differ significantly. According to official sources, Sudan annually used only 12.6 BCM in 1985/86 of the 18.5 BCM to which it is entitled, while according to independent estimates Sudan has already reached this share in the same year.[13]

8.4 The Ethiopian Unknown

The key weakness of the 1959 agreement is, however, that it was unable to include Ethiopia, one of the most important Nile states. When the accord was signed, Ethiopia, under Emperor Haile Selassie, was a close ally of the United States, whereas Egypt, under Nasser, had sought rapprochement with the Soviet Union. And thus it is not by chance that it was precisely at this juncture, i.e. between 1958-63, that the U.S. Federal Bureau of Reclamation conducted a large-scale feasibility study on 33 dams in the catchment area of the Blue Nile, called the Blue Nile Plan. Various sources indicate that the irrigated agriculture made possible in

[11] Abu Zaid and Rady 1992, p.99.

[12] Whittington and Haynes 1985, p.136; Waterbury 1990, pp.284-288; Ayeb 1991, p.67.

[13] Official figures from the PJTC quoted in Chesworth (1994, p.76) state 12.6 BCM/a for 1985/86. Kliot (1994, p.65) estimates, however, up to 21.4 BCM/a for the same year, including 2.5 BCM/a of evaporation between Sennar and Aswan and 0.3 BCM/a of ground water supplies which have to be deducted from that figure.

Ethiopia by these dams would have consumed between 5.4 and 6.4 BCM of water per year.[14] However, this water use could be offset or even be overcompensated by reduced evaporation in Lake Nasser, whose storage volume and area would be reduced, because most of the Nile flood would be stored in the Ethiopian highlands where evaporation is much lower: „Under appropriate working arrangements the amount of water for irrigation throughout the Nile Basin could actually be increased."[15]

With the exception of two smaller dams, this plan was never realized. Yet whenever Ethiopian governments were interested in exerting pressure on Egypt, they have more or less openly pointed to the possibility of building these dams. In 1977 the Marxist Mengistu government came to power in Ethiopia. Egypt, under Sadat, had previously switched over to the Western camp and was supporting Somalia against Ethiopia in the Ogaden War. Ethiopia now once again forwarded its dam plans. Sadat's response was to threaten to have the construction sites bombed if even one dam should be built.[16]

Quite apart from the international conflict potential they entail, it is in any case questionable whether building such a high number of large dams in Ethiopia makes economic sense. The anticipated benefits in electricity generation and agriculture would have to justify the estimated costs of over US$ 50 billion (roughly nine times Ethiopia's gross domestic product). Although it has been pointed out that theoretically „Ethiopia has the possibility to command sufficient water to irrigate a much greater area than currently irrigated in Egypt and Sudan together",[17] Ethiopian officials acknowledge that because of limited or negative economic returns „the Ethiopian authorities are unlikely to invest scarce investment resources to large scale irrigation projects where both initial and running costs are high".[18] What appears economically interesting to Ethiopia is harnessing hydropower with a capacity of 8380 megawatts (the equivalent of eight big nuclear power plants, or four times the capacity of the hydroelectric power plant on the Aswan Dam). This utilization would only slightly reduce the quantity of water (due to greater evaporation in the reservoirs). As the hydropower potential substantially exceeds the expected electricity demand in Ethiopia, „hydropower development plans are based on the assumption that (Ethiopia) will become a major exporter of power".[19] The only potential power importer in the region located relatively close to the dam sites would be Sudan. The Sudanese government, however, has its own ambitious hydropower development plans. But even if the economic viability of the dams were guaranteed, this bitterly poor country would be unable itself to internally finance construction, and would instead be forced to seek grants or credits abroad. It is unlikely that Ethiopia will be able to get development-assistance funds for building all dams originally envisaged or a small number of large dams. Recent years have experienced a substantial growth of skepticism as to the ecological,

---

[14] Ayeb 1991, p.70; Beschorner 1992, p.56.
[15] Collins 1994, p.123.
[16] Beschorner 1992, p.60.
[17] Abate 1994, p.229.
[18] Abate 1994, p.231.
[19] Abate 1994, p.230.

social, and economic costs of large dams. The World Bank, following a period of initial enthusiastic support, has now been prevailed upon to do a spectacular about-face, withdrawing from two vociferously criticized major dam projects (Narmada in India and Arun in Nepal). Also, the World Bank is bound by a directive[20] of its executive board to fund dams on international rivers only if accords on water allocation have been signed with the states downstream of such projects.[21] Many other donors will be reluctant to fund such dams without the World Bank and against the will of Egypt and Sudan.

In June 1996 the Ethiopian government approved two 'micro dams' for hydroelectric power generation on the Blue Nile and the upper Atbara River.[22] If these dams are among the smaller - but still large - of the 33 dams originally envisaged and if they are not used for irrigation, as it seems to be the case, the reduction of Nile water flows will not be high. Nevertheless, the Egyptian government has strongly condemned the proposed dams. It is not yet clear who will finance these dams.

8.5 The Jonglei Canal

Despite a period of frequent changes of government, including numerous coups and even more numerous attempted coups, Sudan has always complied with the 1959 Nile Waters Agreement, including its provisions on the exchange of hydrological data. In 1974 Anwar as-Sadat and Gafaar an-Numeiri agreed, in conformity with the Nile Waters Agreement, to jointly fund a canal through the Sudd; it was intended to increase the flow of the White Nile by some 9 BCM. Including a projected second phase, up to 18.9 BCM per year could be made available.[23] The canal from Bor to Malakal, 360 km in length, was nearly complete when the Sudanese People's Liberation Army, under John Garang, put an end to construction work by abducting, in 1984, the foreign engineers involved in the project.[24] The project then had to be abandoned as a result of the civil war in southern Sudan.

8.6 An Agreement for the Entire Nile Basin?

The UN and the Organization of African Unity (OAU) were at this time at pains to cautiously bring together all of the Nile states in organizations dedicated to technical questions touching on the use of the waters of the Nile. These included the Undugu Working Group, named after the Swahili word for unity, in which, since

[20] Operational Directive 7.50.
[21] World Bank 1993, p.120.
[22] Financial Times, 16 October 1996, p.6.
[23] Howell and Lock (1994) for technical details of the project, and Collins (1994, pp.116-118; 128-134) for its general discussion.
[24] Warburg 1991, p.566.

1982, Egypt, Sudan, and the states on the White Nile have worked together on a cooperative basis. Ethiopia has not participated in the working group, and it has met with little success.[25] Another factor hampering cooperation in the Nile Basin is that, aside from the water issue, the riparian states share few common economic and political interests. There is, for instance, only a minimum of trade between the Nile states.

A modest breakthrough in cooperation on water-related issues was achieved after the fall of the Mengistu government in Ethiopia, when, in December 1991, Egypt and Ethiopia signed an agreement on peace and joint cooperation. The agreement obliges Ethiopia to participate fully in all water-related initiatives in the Nile Basin.[26] This was followed by the series of conferences known as Nile 2002, which began with meetings in Aswan (1993), Khartoum (1994), Arusha/Tanzania (1995), and Kampala/Uganda (1996) and is planned to be continued with annual conferences in the ten Nile states, the next one being scheduled for Addis Ababa in Ethiopia.[27] The participants in the conferences, at which the results of scientific research are presented, include the ministers of the Nile states responsible for water resources management. The conference held in Arusha in February 1995 saw the establishment of an Expert Council which was given the task of working out a framework agreement on the allocation of the waters of the Nile between all the Nile states.[28] An agreement of this sort would constitute an important means for preventing future conflicts over water. While the conference series went on, the Ethiopian minister for environmental protection, Mesfin Abebe, denied his country's reputation of uncooperativeness in issues concerning water when he stated that „the water resources of the Nile can be shared equitably without causing any appreciable harm". The Nile, he noted, is not a source of conflict but a means of regional cooperation.[29]

8.7 Sudan's Position

In contrast to what the political rhetoric would lead one to expect, the recent tensions between Egypt and Sudan at the end of 1995 did not centre on water issues; the actual issue motivating them was the question of Sudanese support for the Islamist opposition in Egypt. In fact, however, Sudan's threats are, in view of its limited potentials, no more than empty words. The relatively small dams in Sudan, i.e. the Roseires Dam and the Sennar Dam on the Blue Nile, the Jebel Auliya Dam on the White Nile, and the Khashm el-Girba Dam on the Atbara, are able to retain only a fraction of the Nile's floodwaters. The overall capacity of the reservoirs, in part already filled by sediment, is estimated today to be at most 6 MCM, some 7%

[25] Wild 1994, p.69.
[26] Beschorner 1992, p.60.
[27] Shady et al. 1994, pp.77-81.
[28] Postel 1995, p.20; The East African 1996.
[29] Abebe 1994.

of the amount of the Nile's annual flow.[30] In view of the poor state of its irrigation systems, Sudan has also for years not been able to completely use the water in the existing reservoirs, to say nothing of using even more water in agriculture. In economic terms, according to foreign agricultural experts, the first priority of Sudan's agriculture should be to rehabilitate its existing irrigation systems.[31] Only then should new irrigation projects be launched.

It has been estimated that six dams for hydroelectric power generation, with a combined installed capacity of 2,760 MW (more than the power plant at the Aswan High dam), could be built on the Main Nile downstream from Khartoum. Total evaporative losses from these dams would amount to almost 6 BCM per year.[32] The largest among these dams is the Hamdab Dam (also called High Merowe Dam) on the fourth cataract.[33] The reservoir behind the 83-metre-high dam would inundate the land presently inhabited by 60,000 people. The evaporation losses in the planned reservoir alone are estimated at some 1.5 BCM per year. The project is planned above all to generate electricity, and so the overall consumptive water use as a result of this dam would not be much higher than this figure.

Other water projects along the Sudanese Nile include the heightening of the Roseires dam to increase the irrigated area in the Gezira scheme, irrigation with direct pumping from the White Nile, construction of a dam on the Setit, i.e. a tributary of the Atbara, for irrigation and increasing irriation intensities on existing schemes.[34] Altogether these projects would use almost an additional 10 BCM of water per year. According to the 1959 treaty, Sudan is entitled only to a small part of this water. For the remaining amount Sudan is placing its hopes in a resumption of the construction of the Jonglei Canal. In view of the political isolation of the Bashir Government, there is at present no prospect of foreign funding for either a rehabilitation or an expansion of the country's irrigation systems. Even though the Sudanese government is currently conducting talks with Chinese and Russian firms on the construction of dams in Sudan,[35] the funding of these billion-dollar projects remains uncertain. Moreover, it is questionable whether the completion of the Jonglei Canal is feasible and whether Western donors would be prepared to fund any new large dams as a result of the anticipated environmental impacts (partial drainage of the Sudd), as well as the social costs (resettlement). This situation exists independently of whether or not the political circumstances in Sudan would permit these projects to be financed.

---

30 Kliot 1994, p.39.

31 Whittington and Haynes 1985, p.134; Kliot 1994, p.70.

32 Knott and Hewett (1994, p.214) quoting the Nile Water Study by Coyne et Bellier, Sir Alexander Gibb & Partners, Hunting Technical Services and Sir M. Macdonald and Partners of 1979.

33 Beck 1996.

34 Knott and Hewett 1994, pp.211-213.

35 Beck 1996, p.2;10.

8.8
Consequences of Climate Change?

An important unknown for Egypt's long-term water supplies is the future development of rainfall in the Nile Basin that will result from the more and more likely prospect of a climate change. The average amount of Nile flow until the end of the 1970's has in fact fallen short of the annual 84 BCM predicted on the basis of the average value for the years 1900-1959.[36] On the other hand, climate models assume that global warming will tend to increase rainfall.[37] Higher temperatures would, however, also increase evaporation. The prognoses on Nile water flows are as yet uncertain with regard to both the direction and the level of the changes in the quantity of Nile waters to be anticipated. Different climate models indicate changes that fluctuate between Nile water quantities increasing by 30% and decreasing by up to 78%.[38] One climate researcher concludes after a detailed analysis that „the current best guess for greenhouse-induced forcing of Nile discharge would be for reduced Blue Nile flows, and constant or slightly increased White Nile flows".[39]

This prognosis, however, seems too uncertain to be used as a basis for future planning.

8.9
Long-Term Risks for Egypt

Egypt is the country with the largest population of the Nile basin, and it has no other substantial water resources other than the Nile. What are Egypt's long-term perspectives in view of the possibility of rising sea levels on one hand, and on the other hand water demand levels set to increase over the long term in Sudan and Ethiopia?

Egypt cannot rule out the possibility that Sudan will one day claim the full water rights to which it is entitled, and that Ethiopia will for the first time obtain and exploit formal Nile water rights that would be defined at the expense of the water rights presently enjoyed by Egypt and Sudan - a hydrological worst-case scenario from the Egyptian perspective. There is at present no discussion underway in Egypt on possible structural reforms that could help to encounter any such long-term development by initiating changes of domestic economic and agricultural policy. This future problem is being shunted in view of the massive political, economic, and social problems presently facing the country.

For Egypt, the consequences of growing water scarcity will be mainly economic in nature: irrigation-based agriculture would have to be offered stronger incentives

---

[36] Waterbury 1979, p.252. This figure neither includes the drought of the mid-80's, nor the exceptionally high flood of 1996.

[37] Deutscher Bundestag 1991, p.60;81.

[38] Magdy et al. 1994.

[39] Hulme 1994, p.159.

to use water more thriftily. This, however, poses difficulties, because in Egypt water for irrigation purposes is traditionally free, and, in addition, the quantity of water consumed is not metred. Some 20 million Egyptians, one third of the country's population, currently live from agriculture.[40]If Sudan and Ethiopia should one day increase their water consumption, Egypt would presumably be unable to reduce the water consumption of its agriculture evenly in all parts of the country.

If the abstraction of water in Upper Egypt and the Cairo region can then not be restricted, the water quality in the Nile Delta will continue to decline. In the delta region Nile water is heavily salinized by return-flows of agricultural drainage water and at the same time polluted by insufficiently treated urban and industrial waste water.[41] Soil salinization resulting from irrigation with this water causes substantially diminished yields and/or drainage costs. Eventually, irrigated areas in the northern Nile Delta might have to be gradually abandoned due to salinization. Furthermore, the quality of drinking water in Alexandria and other cities in or near the Delta will be endangered. The quality of the raw Nile water at the intake of the treatment plants has steadily deteriorated, and some of the plants are therefore unable to comply with Egyptian drinking water standards at all times.[42]

8.10 Expansion of Irrigation in Egypt or Structural Reforms?

Moreover, a large part of the ambitious projects currently being pursued to irrigate the desert would have to be given up. More surprising is that the Egyptian government plans to irrigate 210,000 hectares in the oases of the Western Desert (in the so-called New Delta) with Nile water taken from Lake Nasser. The project, which includes the building of the huge Toshka pumping station at Lake Nasser and the Shaikh Zayed Canal to the area south of the Kharga oasis, was given the official go-ahead at a cabinet meeting in December 1996.[43] The decision was apparently prompted by the exceptionally high Nile floods in autumn 1996, which filled up Lake Nasser for the first time since its construction and caused water to naturally flow in the neighboring Toshka depression.[44]

Regardless of annual fluctuations in the flow of the Nile, and regardless of developments on the upper course of the river, it is inevitable, although not officially acknowledged, that sustainable new employment and income opportunities for the Egyptian population, which is growing by roughly one million every nine months, are very limited in agriculture because of water scarcity. Egypt is therefore forced to invest more intensively in export-oriented industries and the service sector. Even today two thirds of the Egyptian population no longer live from agriculture, and the country is forced to import over two thirds of its food needs.[45] If Egypt

[40] Beschorner 1992, p.48.
[41] Abu Zeid and Rady 1992, p.97.
[42] Institut Méditerranéen de l'Eau 1995, pp.26.
[43] Middle East Economic Digest, January 1997.
[44] Financial Times, 16 October 1996, p.6.
[45] Beschorner 1992, p.51.

does not succeed in creating employment outside agriculture, even more Egyptians will have to emigrate to other Arab countries, e.g. Libya, Iraq, Saudi Arabia, in search of work, provided these countries are willing to accept them.

8.11 Nile Water for the Sinai, And for Gaza and Israel?

The recently completed Salam Canal conducts Nile water beneath the Suez Canal into the Sinai for irrigation as far as El Arish on the border to the Gaza Strip. The economic viability of the project and the long-term availability of surplus Nile waters are doubtful. Nevertheless, the extension of the canal to the Gaza strip and even to Israel has been suggested. One study found that by itself conversion of irrigated agriculture in the Sinai to more economical drip irrigation would save enough Nile water to cover the water needs of the densely populated Gaza Strip without much additional effort and costs.[46]

A canal carrying Nile waters to Israel had been anticipated as early as 1902 by Theodor Herzl and was proposed again by Anwar as-Sadat after the Camp David peace treaty. The proposal has, however, met with vehement protest on the part of the Islamist opposition and nationalist circles in Egypt.[47] This proposal will thus, for political reasons, without doubt not be realized.

8.12 Genuine Water Conflicts or Political Conflicts?

Egyptian fears of substantially reduced inflows of water as a result of irrigation projects on the upper course of the Nile are largely without foundation in the immediate future. Sudan and Ethiopia currently do not have the capacity to build the large dams required for such projects.

For the decades to come, however, everything depends on whether the upstream countries will significantly expand irrigated agriculture or not. If this is the case, water inflows to Egypt will be reduced and a genuine water conflict will occur. Judging from past experience, the building of any dam, even if exclusively used for hydropower generation, will cause sharp Egyptian protests. But from a hydrological point of view - with the possible exception of the Hamdab dam in Northern Sudan with its high evaporative losses - these concerns are unfounded.

Up today, there has been no genuine water conflict in the Nile Basin. It is perhaps not by chance that the, at times vehement, rhetoric about water between Egypt, Sudan, and Ethiopia flares up whenever other conflicts enter a particularly acute phase. In the 1980's, for instance, when relations between Sudan and Egypt were good, water was not the object of public verbal attacks. Unlike at the end of the 1950's and of the 1970's, water is today to a much lesser degree an object of conflict between Egypt and Ethiopia, in spite of the much higher population. This is, however, not to rule out the possibility that over the long run genuine water

[46] Dinar and Wolf 1994, pp.43-62.
[47] Hottinger 1992, pp.153-163.

conflicts might emerge in the Nile Basin, if irrigated agriculture in the upstream countries was actually to be significantly expanded.

8.13 Regional Water Interests of the Nile States

Taking the above analysis into account, it is in Egypt's long-term interest to curtail the irrigated agriculture in Sudan and Ethiopia. Yet Egypt's foreign policy could, in terms of water policy, prove to be ambivalent. On one hand, a stable and pro-Egyptian government in Sudan might resume construction of the Jonglei Canal, which would be in Egypt's interest. On the other hand, stable conditions in Sudan and Ethiopia might lead to a realization of the dams planned there. It can therefore not be ruled out that persistent instability in Sudan and in Ethiopia might lie in Egypt's interest, i.e. as a means of ensuring a continuation of the Nile's present flow.

Should they one day have the financial resources required, Sudan and Ethiopia, on the other hand, would be interested in increasing their water use. They might then attempt to play the 'water card' as a trump against Egypt with an eye to achieving their interests in other areas. But before any such genuine water conflicts go beyond the level of mere rhetoric, quite a bit of water will have flowed down the Nile.

References

Abate, Z. 1994: The Integrated Development of Nile Waters, in: P. P. Howell; J. A. Allan (eds.), pp.227-241.

Abebe, Mesfin 1994: The Nile - Source of Regional Cooperation or Conflict?, VIII IWRA-Congress, Kairo, 24 November.

Abu Zeid, M.A.; Rady, M.A. 1992: Water Resources Management and Policies in Egypt, in: Country Experiences with Water Resources Management, World Bank Technical Paper No.175, Washington D.C.

Ayeb, Habib 1992: La vallée du Nil. Un grand axe géopolitique, in: Monde arabe Maghreb Machrek. No.138, Oct.-Déc.

Beck, Kurt 1996: Wer kennt schon Hamdab? Ein Staudammvorhaben im Sudan (Who knows Hamdab? A proposed Dam Project in Sudan). unpublished manuscript, Bayreuth.

Beschorner, Natascha 1992: Water and Instability in the Middle East, Adelphi Paper 273, International Institute for Strategic Studies, London.

Collins, R. 1994: History, Hydropolitics and the Nile: Myth or Reality, in: P. P. Howell; J. A. Allan (eds.), pp.109-135.

Deutscher Bundestag 1991: Erster Bericht der Enquete-Kommission 'Schutz der Erdatmosphäre' zum Thema Klimaänderung gefährdet globale Entwicklung (First Report of the Commission 'Protection of the Earth's Atmosphere' on the subject Climate Change endangers Global Development), pp.60-81, Bonn.

Dinar, Ariel; Wolf, Aaron 1994: International Markets for Water and the Potential for Regional Cooperation: Economic and Political Perspectives in the Western Middle East, in: Economic Development and Cultural Change, pp.43-62.

Financial Times, 16 October 1996
Hottinger, A. 1992: Wasser als Konfliktstoff. Eine Existenzfrage für Staaten des Nahen Ostens (Water as a Source of Conflict. A Vital Issue for the Middle East States), in: Europa-Archiv, 1992/6, pp.153-163.
Howell, P.; Lock, M. 1994: The Control of the Swamps of the Southern Sudan: Drainage schemes, loal effects and environmental constraints on remedial development, in: P. P. Howell; J. A. Allan (eds.), pp.243-279.
Howell, P.P.; J. A. Allan (eds.) 1994: The Nile. Sharing a Scarce Resource. A Historical and Technical Review of Water Management and of Economic and Legal Issues, Cambridge.
Hulme, M. 1994: Global Cimate Change and the Nile Basin, in: P. P. Howell; J. A. Allan (eds.), pp.139-162.
Kabanda, B.; Kahingere, P. 1994: Irrigation and Hdropower Potential. Water Needs in Uganda, and Overview, in: P. P. Howell; J. A. Allan (eds.), pp.217-226.
Kliot, Nurit 1994: Water Resources and Conflict in the Middle East. London, New York.
Knott, D.; Hewitt, R. 1994: Future Water Development Planning in the Sudan, in: P. P. Howell; J. A. Allan (eds.), pp.205-216.
Magdy; Strzepek; Yates 1994: Potential Climate Change Impacts on Nile Basin, Proceedings of the VIII IWRA Congress on Water Resources, Cairo, November 21-25, Volume 1, T4-S2.
Middle East Economic Digest, January 1997
Okidi, C.O. 1994: History of the Nile Basin and Lake Victoria Basins Through Treaties, in: P. P. Howell; J. A. Allan (eds.), pp.321-350.
Postel, Sandra 1995: Sag mir wo die Flüsse sind! (Where have all the Rivers Gone), World-Watch July/August.
Shady, Aly M.; Adam, Ahmad M; Mohamed, Kamal Ali 1994: The Nile 2002: The Vision Toward Cooperation in the Nile Basin, in: Water International, 19, pp.77-81.
Stoner, R. 1994: Future Irrigation Planning in Egypt, in: P. P. Howell; J. A. Allan (eds.), pp.195-204.
The East African, Nairobi, quoted in: Aktueller Informationsdienst Afrika, 22 March 1996.
The World Bank 1993: Water Resources Management. A World Bank Policy Paper, Washington D.C.
United Nations 1963: Agreement for the Full Utilization of the Nile Waters. Signed at Cairo, on 8 November 1959, Treaty Series 543, pp.63-77.
Warburg, G. R. 1991: The Nile in Egyptian-Sudanese Relations, in: Orient 32/4, pp.565-572.
Waterbury , John 1979: Hydropolitics of the Nile Valley, Syracuse.
Waterbury, John 1990: Legal and Institutional Arrangements for Managing Water Resources in the Nile Basin, in: Simon, Reeva S. (ed.), The Middle East and North Africa. New York.
Whittington, Dale; Haynes, Kingsley E. 1985: Nile Water for Whom?, in: P. Beaumont; K. McLachlan, Agricultural Development in the Middle East.
Wild, J. 1994: Towards an Economic Reallocation of Blue Nile Waters, unpublished manuscript, London.

Annex[1]

Convention on the Law of the Non-navigational Uses of International Watercourses (21 May 1997)

(Articles reproduced by extract from United Nations Document A/51/869)[2]

Preamble

The Parties to the present Convention,

Conscious of the importance of international watercourses and the non-navigational uses thereof in many regions of the world,

Having in mind Article 13, paragraph 1 (a), of the Charter of the United Nations, which provides that the General Assembly shall initiate studies and make recommendations for the purpose of encouraging the progressive development of international law and its codification,

Considering that successful codification and progressive development of rules of international law regarding non-navigational uses of international watercourses would assist in promoting and implementing the purposes and principles set forth in Articles 1 and 2 of the Charter of the United Nations,

Taking into account the problems affecting many international watercourses resulting from, among other things, increasing demands and pollution,

Expressing the conviction that a framework convention will ensure the utilization, development, conservation, management and protection of international watercourses and the promotion of the optimal and sustainable utilization thereof for present and future generations,

Affirming the importance of international cooperation and goodneighbourliness in this field,

Aware of the special situation and needs of developing countries,

Recalling the principles and recommendations adopted by the United Nations Conference on Environment and Development of 1992 in the Rio Declaration and Agenda 21,

---

1 For the Turkish-Syrian Protocol on Economic Cooperation, signed in Damascus on 17 July 1987, see Box 7.1., p.122.

2 The text of the Convention, the Annex Arbitration and the Statements of Understanding are reproduced in: International Legal Materials, 3/1997, pp.700-720 (36 I.L.M. 700(1997)).
For additional information contact the UN Treaty Section, Office of Legal Affairs, UN Headquarters, New York, NY 10017 U.S.A., tel.: (212) 963-5047, fax: ...-3693.

Recalling also the existing bilateral and multilateral agreements regarding the non-navigational uses of international watercourses,

Mindful of the valuable contribution of international organizations, both governmental and non-governmental, to the codification and progressive development of international law in this field,

Appreciative of the work carried out by the International Law Commission on the law of the non-navigational uses of international watercourses,

Bearing in mind United Nations General Assembly resolution 49/52 of 9 December 1994,

Have agreed as follows:

Part I
Introduction

Article 1
Scope of the Present Convention[3]

1. The present Convention applies to uses of international watercourses and of their waters for purposes other than navigation and to measures of protection, preservation and management related to the uses of those watercourses and their waters.
2. The uses of international watercourses for navigation is not within the scope of the present Convention except insofar as other uses affect navigation or are affected by navigation.

---

[3] Discussion in the Sixth Committee: *France*: "Article 1 must also convey the idea that a framework convention had no direct effect in law" (A/C.6/51/SR.12, p.6). *China*: "...article 1 should clarify that the framework convention was not to intended to supplant existing or future bilateral or multilateral watercourse agreements" (p.9).

Article 2
Use of Terms[4]

For the purposes of the present Convention:

a) 'Watercourse' means a system of surface waters and groundwaters constituting by virtue of their physical relationship a unitary whole and normally flowing into a common terminus;
b) 'International watercourse' means a watercourse, parts of which are situated in different States;
c) 'Watercourse State' means a State Party to the present Convention in whose territory part of an international watercourse is situated, or a Party that is a regional economic integration organization, in the territory of one or more of whose Member States part of an international watercourse is situated;
d) 'Regional economic integration organization' means an organization constituted by sovereign States of a given region, to which its member States have transferred competence in respect of matters governed by this Convention and which has been duly authorized in accordance with its internal procedures, to sign, ratify, accept, approve or accede to it.

---

4 Discussion in the Sixth Committee: The definition of 'international watercourse' and the question as to whether it would include the groundwater connected with a watercourse was controversial. Turkey and other states made reservations. Report of the Secretary-General: Comments and Observations from States, A/51/275, 6 August 1996: "*Ethiopia* does not agree with the formulation of draft article 2 (b) in its present form as it applies to all the hydrographic components of international watercourse. We hold that watercourse should be treated as having a "relative international character ... This could result in excessive interference by States in each other's legitimate internal affairs" (p.22). *Turkey* insists on differentiation between watercourses forming a boundary and transboundary watercourses: "... to make no distinction between watercourses forming a boundary and transboundary watercourses and then subject these two categories to the same legal rules is unrealistic as well as against the legal principle of equitable utilization" (p.24), and proposes as Proposal in CRP 60: "1. Add the following sentence to subparagraph (a): International watercourses are divided into the following two categories: Watercourses which form a boundary; Transboundary watercourses which flow successively through two or more States..." Eight downstream states demanded that a definition of the term 'optimal utilization' be included (L.1.Add.4). This, too, was a matter of sharp dispute.

Article 3
Watercourse Agreements[5]

1. In the absence of an agreement to the contrary, nothing in the present Convention shall affect the rights or obligations of a watercourse State arising from agreements in force for it on the date on which it became a party to the present Convention.
2. Notwithstanding the provisions of paragraph 1, parties to agreements referred to in paragraph 1 may, where necessary, consider harmonizing such agreements with the basic principles of the present Convention.
3. Watercourse States may enter into one or more agreements, hereinafter referred to as 'watercourse agreements', which apply and adjust the provisions of the present Convention to the characteristics and uses of a particular international watercourse or part thereof.
4. Where a watercourse agreement is concluded between two or more watercourse States, it shall define the waters to which it applies. Such an agreement may be entered into with respect to an entire international watercourse or any part thereof or a particular project, programme or use except insofar as the agreement adversely affects, to a significant extent, the use by one or more other watercourse States of the waters of the watercourse, without their express consent.
5. Where a watercourse State considers that adjustment and application of the provisions of the present articles is required because of the characteristics and uses of a particular international watercourse, watercourse States shall consult with a view to negotiating in good faith for the purpose of concluding a watercourse agreement or agreements.
6. Where some but not all watercourse States to a particular international watercourse are parties to an agreement, nothing in such an agreement shall affect the rights or obligations under the present Convention of watercourse States that are not parties to such an agreement.

---

[5] Adopted by vote: yes: 36, no: 3 (France, Turkey, Ethiopia), abstain: 21 (among them: China, Israel, Jordan).
Statement of understanding noted by the Chairman of the Working Group (WG/L.4/Add.1): "(a) The present Convention will serve as a guideline for future watercourse agreements and, once such agreements are concluded, it will not alter the rights and obligations provided therein, unless such agreements provide otherwise. (b) The term 'significant' is not used in the sense of 'substantial'. What is to avoided are localized agreements, or agreements concerning a particular project, programme or use, which have a significant adverse effect upon third watercourse State. While such an effect must be capable of being established by objective evidence and not be trivial in nature, it need not rise to the level of being substantial."
Discussion in the Sixth Committee: There was a direct link between the discussion on article 1 and 3. *France*: "The discussion ... demonstrated that interpretation of article 4 and of the following articles was closely linked to the interpretation to be given to article 3, namely the scope to be given to the framework convention" (A/C.6/51/SR.13, p.11).

Article 4
Parties to Watercourse Agreements

1. Every watercourse State is entitled to participate in the negotiation of and to become a party to any watercourse agreement that applies to the entire international watercourse, as well as to participate in any relevant consultations.
2. A watercourse State whose use of an international watercourse may be affected to a significant extent by the implementation of a proposed watercourse agreement that applies only to a part of the watercourse or to a particular project, programme or use is entitled to participate in consultations on such an agreement and, where appropriate in the negotiation thereof in good faith with a view to becoming a party thereto, to the extent that its use is thereby affected.

Article 5
Equitable and Reasonable Utilization and Participation[6]

1. Watercourse States shall in their respective territories utilize an international watercourse in an equitable and reasonable manner. In particular, an international watercourse shall be used and developed by watercourse States with a view to attaining optimal and sustainable utilization thereof and benefits therefrom, taking into account the interests of the watercourse States concerned, consistent with adequate protection of the watercourse.
2. Watercourse States shall participate in the use, development and protection of an international watercourse in an equitable and reasonable manner. Such participation includes both the right to utilize the watercourse and the duty to co-

[6] The articles 5, 6, 7 were adopted by vote as a package: yes: 38, no: 4 (China, France, Turkey, Tanzania), abstain: 22 (among them: Egypt, Ethiopia, India, Pakistan, Sudan).
Discussion in the Sixth Committee: The different opinions about equitable and reasonable utilization could not be compromised on. Moreover, consensus could not be reached on 'sustainable utilization' and on an extension: "...protection of the watercourse and its ecosystem". Several delegations made reservations. *Turkey*: "It is necessary that equitable and reasonable utilization should be understood and interpreted in the light of the fundamental principle of the sovereign rights of States over their territory. It should also be applied by taking fully into account all the particularities of the watercourses, including the distinction of whether they are transboundary by nature or form a boundary between States. ...optimal utilization should not be restricted to protection only but should be seen also as comprising the concept of efficient use" (A/51/275, p.36). *Ethiopia*: "...basic consideration should be given to equitable and reasonable use in relation to harm. If watercourse States adhere to the principle of equitable and reasonable use, the danger of causing significant harm to an international watercourse will thereby be eliminated. Equitable and reasonable use should, therefore, be the overriding consideration and significant harm should be subordinated to it" (A/51/275, p.33f). *Switzerland*: "If more importance was assigned to article 7 than to article 5 ... downstream States would benefit at the expense of upstream States ... That being the case ... that article 7 should simply be deleted" (A/C.6/51/SR.15, p.5). *USA*:"...we will need to consider not only the protection of watercourses, but also the protection of associated ecosystems, including the coastal ecosystems into which most watercourses flow" (A/51/275, p.37).

operate in the protection and development thereof, as provided in the present Convention.

Article 6
Factors Relevant to Equitable and Reasonable Utilization

1. Utilization of an international watercourse in an equitable and reasonable manner within the meaning of article 5 requires taking into account all relevant factors and circumstances, including:
   a) geographic, hydrographic, hydrological, climatic, ecological and other factors of a natural character;
   b) the social and economic needs of the watercourse States concerned;
   c) the population dependent on the watercourse in each watercourse State;
   d) the effects of the use or uses of the watercourse in one watercourse State on other watercourse States;
   e) existing and potential uses of the watercourse;[7]
   f) conservation, protection, development and economy of use of the water resources of the watercourse and the costs of measures taken to that effect;
   g) the availability of alternatives, of corresponding value, to a particular planned or existing use.

2. In the application of article 5 or paragraph 1 of this article, watercourse States concerned shall, when the need arises, enter into consultations in a spirit of cooperation.
3. The weight to be given to each factor is to be determined by its importance in comparison with that of other relevant factors. In determining what is a reasonable and equitable use, all relevant factors are to be considered together and a conclusion reached on the basis of the whole.

Article 7
Obligation not to Cause Significant Harm

1. Watercourse States shall, in utilizing an international watercourse in their territories, take all appropriate measures to prevent the causing of significant harm to other watercourse States.
2. Where significant harm nevertheless is caused to another watercourse State, the States whose use causes such harm shall, in the absence of agreement to such use, take all appropriate measures, having due regard for the provisions of articles 5 and 6, in consultation with the affected State, to eliminate or mitigate such harm and, where appropriate, to discuss the question of compensation.

---

[7] Statement of understanding noted by the Chairman of the Working Group (WG/L.4/Add.1): "In order to determine whether a particular use is equitable and reasonable, the benefits as well as the negative consequences of a particular use should be taken into account."

Part II
General Principles

Article 8
General Obligation to Cooperate

1. Watercourse States shall cooperate on the basis of sovereign equality, territorial integrity, mutual benefit and good faith in order to attain optimal utilization and adequate protection of an international watercourse.
2. In determining the manner of such cooperation, watercourse States may consider the establishment of joint mechanisms or commissions, as deemed necessary by them, to facilitate cooperation on relevant measures and procedures in the light of experience gained through cooperation in existing joint mechanisms and commissions in various regions.

Article 9
Regular Exchange of Data and Information

1. Pursuant to article 8, watercourse States shall on a regular basis exchange readily available data and information on the condition of the watercourse, in particular that of a hydrological, meteorological, hydrogeological and ecological nature and related to the water quality as well as related forecasts.
2. If a watercourse State is requested by another watercourse State to provide data or information that is not readily available, it shall employ its best efforts to comply with the request but may condition its compliance upon payment by the requesting State of the reasonable costs of collecting and, where appropriate, processing such data or information.
3. Watercourse States shall employ their best efforts to collect and, where appropriate, to process data and information in a manner which facilitates its utilization by the other watercourse States to which it is communicated.

Article 10
Relationship Between Different Kinds of Uses[8]

1. In the absence of agreement or custom to the contrary, no use of an international watercourse enjoys inherent priority over other uses.
2. In the event of a conflict between uses of an international watercourse, it shall be resolved with reference to articles 5 to 7, with special regard being given to the requirements of vital human needs.

[8] Statement of understanding noted by the Chairman of the Working Group (WG/L.4/Add.1): "In determining 'vital human needs', special attention is to be paid to providing sufficient water to sustain human life, including both drinking water and water required for production of food in order to prevent starvation."

Part III
Planned Measures

Article 11
Information Concerning Planned Measures[9]

Watercourse States shall exchange information and consult each other and, if necessary, negotiate on the possible effects of planned measures on the condition of an international watercourse.

Article 12
Notification Concerning Planned Measures with Possible Adverse Effects

Before a watercourse State implements or permits the implementation of planned measures which may have a significant adverse effect upon other watercourse States, it shall provide those States with timely notification thereof. Such notification shall be accompanied by available technical data and information, including the results of any environmental impact assessment, in order to enable the notified States to evaluate the possible effects of the planned measures.

Article 13
Period for Reply to Notification

Unless otherwise agreed:

a) A watercourse State providing a notification under article 12 shall allow the notified States a period of six months within which to study and evaluate the possible effects of the planned measures and to communicate the findings to it;
b) This period shall, at the request of a notified State for which the evaluation of the planned measures poses special difficulty, be extended for a period of six months.

[9] Discussion in the Sixth Committee: Due to the obligation to notify planned measures, articles 11 through 19 were altogether not acceptable for Ethiopia and Turkey. Turkey presented own drafts for articles 12 through 15 which were also approved by Ethiopia. *Turkey*: "Consequently, the dispositions of part III should be reduced to a minimum necessary to set forth certain general principles regarding planned measures ... it is considered that the provision of exchange of information and consultations only in the case of water utilization liable to cause significant harm would be sufficient in respect of satisfying the criteria of equitable and reasonable utilization of international watercourses" (A/51/275, p.53). *Ethiopia*: "This article in combination with article 13 to 18, which set out procedures for notification, does not do justice to States that have not developed their water resources vis-à-vis those that have already done so. If a watercourse State has already developed its part of a watercourse, it can halt the other States that have not yet developed theirs by the application of the provisions of this article and articles 13 through 18" (A/51/275, p.54).

Article 14
Obligations of the Notifying State During the Period for Reply

During the period referred to in article 13, the notifying State:

a) Shall cooperate with the notified States by providing them, on request, with any additional data and information that is available and necessary for an accurate evaluation; and
b) Shall not implement or permit the implementation of the planned measures without the consent of the notified States.

Article 15
Reply to Notification

The notified States shall communicate their findings to the notifying State as early as possible within the period applicable pursuant to article 13. If a notified State finds that implementation of the planned measures would be inconsistent with the provisions of articles 5 or 7, it shall attach to its finding a documented explanation setting forth the reasons for the finding.

Article 16
Absence of Reply to Notification

1. If, within the period applicable pursuant to article 13, the notifying State receives no communication under article 15, it may, subject to its obligations under articles 5 and 7, proceed with the implementation of the planned measures, in accordance with the notification and any other data and information provided to the notified States.
2. Any claim to compensation by a notified State which has failed to reply within the period applicable pursuant to article 13 may be offset by the costs incurred by the notifying State for action undertaken after the expiration of the time for a reply which would not have been undertaken if the notified State had objected within that period.

Article 17
Consultations and Negotiations Concerning Planned Measures

1. If a communication is made under article 15 that implementation of the planned measures would be inconsistent with the provisions of articles 5 or 7, the notifying State and the State making the communication shall enter into consultations and, if necessary, negotiations with a view to arriving at an equitable resolution of the situation.
2. The consultations and negotiations shall be conducted on the basis that each State must in good faith pay reasonable regard to the rights and legitimate interests of the other State.
3. During the course of the consultations and negotiations, the notifying State shall, if so requested by the notified State at the time it makes the communication, refrain from implementing or permitting the implementation of the planned measures for a period of six months unless otherwise agreed.

Article 18
Procedures in the Absence of Notification

1. If a watercourse State has reasonable grounds to believe that another watercourse State is planning measures that may have a significant adverse effect upon it, the former State may request the latter to apply the provisions of article 12. The request shall be accompanied by a documented explanation setting forth its grounds.
2. In the event that the State planning the measures nevertheless finds that it is not under an obligation to provide a notification under article 12, it shall so inform the other State, providing a documented explanation setting forth the reasons for such finding. If this finding does not satisfy the other State, the two States shall, at the request of that other State, promptly enter into consultations and negotiations in the manner indicated in paragraphs 1 and 2 of article 17.
3. During the course of the consultations and negotiations, the State planning the measures shall, if so requested by the other State at the time it requests the initiation of consultations and negotiations, refrain from implementing or permitting the implementation of those measures for a period of six months unless otherwise agreed.

Article 19
Urgent Implementation of Planned Measures

1. In the event that the implementation of planned measures is of the utmost urgency in order to protect public health, public safety or other equally important interests, the State planning the measures may, subject to articles 5 and 7, immediately proceed to implementation, notwithstanding the provisions of article 14 and paragraph 3 of article 17.
2. In such case, a formal declaration of the urgency of the measures shall be communicated without delay to the other watercourse States referred to in article 12 together with the relevant data and information.
3. The State planning the measures shall, at the request of any of the States referred to in paragraph 2, promptly enter into consultations and negotiations with it in the manner indicated in paragraphs 1 and 2 of article 17.

Part IV
Protection, Preservation and Management

Article 20
Protection and Preservation of Ecosystems[10]

Watercourse States shall, individually and, where appropriate, jointly, protect and preserve the ecosystems of international watercourses.

Article 21
Prevention, Reduction and Control of Pollution[11]

1. For the purpose of this article, 'pollution of an international watercourse' means any detrimental alteration in the composition or quality of the waters of an international watercourse which results directly or indirectly from human conduct.
2. Watercourse States shall, individually and, where appropriate, jointly, prevent, reduce and control the pollution of an international watercourse that may cause significant harm to other watercourse States or to their environment, including harm to human health or safety, to the use of the waters for any beneficial purpose or to the living resources of the watercourse. Watercourse States shall take steps to harmonize their policies in this connection.
3. Watercourse States shall, at the request of any of them, consult with a view to arriving at mutually agreeable measures and methods to prevent, reduce and control pollution of an international watercourse, such as:
   a) Setting joint water quality objectives and criteria;
   b) Establishing techniques and practices to address pollution from point and non-point sources;
   c) Establishing lists of substances the introduction of which into the waters of an international watercourse is to be prohibited, limited, investigated or monitored.

[10] Discussion in the Sixth Committee: *China* proposed to replace 'preserve the ecosystem' by 'maintain the ecological balance'. This proposal was supported by many states, as the current formulation could be misunderstood. *Ethiopia*: "Preservation could mean that the existing ecosystem, whether good or bad, should be maintained" (A/51/275, p.59).

[11] As regards articles 21, 22, 23 statement of understanding noted by the Chairman of the Working Group (WG/L.4/Add.1): "As reflected in the commentary of the International Law Commission, these articles impose a due diligence standard on watercourse States."
Discussion in the Sixth Committee: The dissent regarding article 21 para. 3 extends from the demand of complete deletion to the inclusion of a detailed list of different measures aimed at pollution control. *Hungary*: "Article 21 of the 1994 draft articles should incorporate the polluter-pays-principle by virtue of which costs of pollution prevention, reduction and control measures shall be borne by the polluter...the concepts of best available technology and best environmental practices introduced and defined by the 1992 Helsinki Convention on the Protection and Use of Transboundary Watercourses and Lakes should be incorporated with respect to the specific obligations incumbent on riparian States" (A/51/275, p.61).

Article 22
Introduction of Alien or New Species

Watercourse States shall take all measures necessary to prevent the introduction of species, alien or new, into an international watercourse which may have effects detrimental to the ecosystem of the watercourse resulting in significant harm to other watercourse States.

Article 23
Protection and Preservation of the Marine Environment

Watercourse States shall, individually and, where appropriate, in cooperation with other States, take all measures with respect to an international watercourse that are necessary to protect and preserve the marine environment, including estuaries, taking into account generally accepted international rules and standards.

Article 24
Management

1. Watercourse States shall, at the request of any of them, enter into consultations concerning the management of an international watercourse, which may include the establishment of a joint management mechanism.
2. For the purposes of this article, 'management' refers, in particular, to:
   a) Planning the sustainable development of an international watercourse and providing for the implementation of any plans adopted; and
   b) Otherwise promoting the rational and optimal utilization, protection and control of the watercourse.

Article 25
Regulation

1. Watercourse States shall cooperate, where appropriate, to respond to needs or opportunities for regulation of the flow of the waters of an international watercourse.
2. Unless otherwise agreed, watercourse States shall participate on an equitable basis in the construction and maintenance or defrayal of the costs of such regulation works as they may have agreed to undertake.
3. For the purposes of this article, 'regulation' means the use of hydraulic works or any other continuing measure to alter, vary or otherwise control the flow of the waters of an international watercourse.

Article 26
Installations

1. Watercourse States shall, within their respective territories, employ their best efforts to maintain and protect installations, facilities and other works related to an international watercourse.

2. Watercourse States shall, at the request of any of them which has reasonable grounds to believe that it may suffer significant adverse effects, enter into consultations with regard to:
   a) The safe operation and maintenance of installations, facilities or other works related to an international watercourse; and
   b) The protection of installations, facilities or other works from wilful or negligent acts or the forces of nature.

Part V
Harmful Conditions and Emergency Situations

Article 27
Prevention and Mitigation of Harmful Conditions

Watercourse States shall, individually and, where appropriate, jointly, take all appropriate measures to prevent or mitigate conditions related to an international watercourse that may be harmful to other watercourse States, whether resulting from natural causes or human conduct, such as flood or ice conditions, water-borne diseases, siltation, erosion, salt-water intrusion, drought or desertification.

Article 28
Emergency Situations

1. For the purposes of this article, 'emergency' means a situation that causes, or poses an imminent threat of causing, serious harm to watercourse States or other States and that results suddenly from natural causes, such as floods, the breaking up of ice, landslides or earthquakes, or from human conduct, such as industrial accidents.
2. A watercourse State shall, without delay and by most expeditious means available, notify other potentially affected States and competent international organizations of any emergency originating within its territory.
3. A watercourse State within whose territory an emergency originates shall, in cooperation with potentially affected States and, where appropriate, competent international organizations, immediately take all practicable measures necessitated by the circumstances to prevent, mitigate and eliminate harmful effects of the emergency.
4. When necessary, watercourse States shall jointly develop contingency plans for responding to emergencies, in cooperation, where appropriate, with other potentially affected States and competent international organizations.

Part VI
Miscellaneous Provisions

Article 29
International Watercourses and Installations in Time of Armed Conflict

International watercourses and related installations, facilities and other works shall enjoy the protection accorded by the principles and rules of international law applicable in international and non-international armed conflict and shall not be used in violation of those principles and rules.

Article 30
Indirect Procedures

In cases where there are serious obstacles to direct contacts between watercourse States, the States concerned shall fulfill their obligations of cooperation provided for in the present Convention, including exchange of data and information, notification, communication, consultations and negotiations, through any indirect procedure accepted by them.

Article 31
Data and Information Vital to National Defence or Security

Nothing in the present Convention obliges a watercourse State to provide data or information vital to its national defence or security. Nevertheless, that State shall cooperate in good faith with the other watercourse States with a view to providing as much information as possible under the circumstances.

Article 32
Non-discrimination[12]

Unless the watercourse States concerned have agreed otherwise for the protection of the interests of persons, natural or juridical, who have suffered or are under a serious threat of suffering significant transboundary harm as a result of activities related to an international watercourse, a watercourse State shall not discriminate on the basis of nationality or residence or place where the injury occurred, in

[12] Discussion in the Sixth Committee: The conflict as to whether and to which extent the principle that native citizens are to be treated equally should also apply to foreigners was sparked off by the principle of non-discrimination. A Russian proposal for a compromise merely concedes a formal guarantee of legal recourse - equal rights regarding the access to judicial procedures - without granting substantive rights. *Russia*: "The watercourse States shall not discriminate on the basis of nationality or residence in granting access to judicial or other procedures, in accordance with their legal systems, to any natural or juridical person who has suffered or is under threat of suffering significant harm as a result of activities related to an international watercourse" (A/C.6/51/NUW/WG/L.1/Add.3, p.2). India, Columbia, Tanzania and Uganda completely refuse to have this principle laid down in adjective as well as in substantive law.

granting to such persons, in accordance with its legal system, access to judicial or other procedures, or a right to claim compensation or other relief in respect of significant harm caused by such activities carried on in its territory.

Article 33
Settlement of Disputes[13]

1. In the event of a dispute between two or more Parties concerning the interpretation or application of the present Convention, the Parties concerned shall seek a settlement of the dispute by peaceful means in accordance with the following provisions.
2. If the Parties concerned cannot reach agreement by negotiation requested by one of them, they may jointly seek the good offices of, or request mediation or conciliation by, a third party, or make use, as appropriate, of any joint watercourse institutions that may have been established by them or agree to submit the dispute to arbitration or to the International Court of Justice.
3. Subject to the operation of paragraph 10, if after six months from the time of the request for negotiations referred to in paragraph 2, the Parties concerned have not been able to settle their dispute through negotiation or any other means referred to in paragraph 2, the dispute shall be submitted, at the request of any of the Parties to the dispute, to impartial fact-finding in accordance with paragraphs 4 to 9, unless the Parties otherwise agree.
4. A Fact-Finding Commission shall be established, composed of one member nominated by each party concerned and in addition a member not having the nationality of any of the parties concerned chosen by the nominated members who shall serve as Chairman.

---

[13] Adopted by vote: yes: 33, no: 5 (China, Colombia, France, India, Turkey), abstain: 25 (among them: Egypt, Ethiopia, Israel, Jordan, Pakistan, Sudan, Syria).
Discussion in the Sixth Committee: Four proposals regarding this draft have been put forth (by Finland, Greece and Italy; Guatemala; (CRP.62/Rev.1, CRP.71) Syria; and by the Chairman of the Drafting Committee) whereby only Syria's proposal contained a compulsory settlement of disputes in the strict sense. *Turkey*: "It would be more appropriate not to foresee any compulsory rules as regards the settlement of disputes, and to leave this issue to the discretion of the States concerned....A framework agreement should not attempt to set forth detailed rules in this respect, since it is virtually impossible to respond to the exigencies of specific and more often than not complex cases of water disputes. Therefore, article 33 should either be omitted or replaced by a general provision on settlement of disputes" (A/51/275, p.68). *France*: "... article 33 was too long and detailed for a framework convention ... the International Law Commission ... should have concentrated only on the essential principles thereof" (A/C.6/51/SR.21, p.7). *Syria*: "...(c) If after six months from the initial request for fact-finding mediation or conciliation, or three months after receipt of a report from the Commission, whichever is the later, the States concerned have been unable to settle the dispute, they shall by agreement submit the dispute to arbitration and if they fail within two months to agree on the arbitration procedure any of them is entitled to submit the dispute to the International Court of Justice" (Proposal by the Syrian Arab Republic for the Article 33, 24 Oct. 1996).

5. If the members nominated by the Parties are unable to agree on a Chairman within three months of the request for the establishment of the Commission, any Party concerned may request the Secretary-General of the United Nations to appoint the Chairman who shall not have the nationality of any of the parties to the dispute or of any riparian State of the watercourse concerned. If one of the Parties fails to nominate a member within three months of the initial request pursuant to paragraph 3, any other Party concerned may request the Secretary-General of the United Nations to appoint a person who shall not have the nationality of any of the Parties to the dispute or of any riparian State of the watercourse concerned. The person so appointed shall constitute a single-member Commission.
6. The Commission shall determine its own procedure.
7. The Parties concerned have the obligation to provide the Commission with such information as it may require and, on request, to permit the Commission to have access to their respective territory and to inspect any facilities, plant, equipment, construction or natural feature relevant for the purpose of its inquiry.
8. The Commission shall adopt its report by a majority vote, unless it is a single-member Commission, and shall submit that report to the Parties concerned setting forth its findings and the reasons therefor and such recommendations as it deems appropriate for an equitable solution of the dispute, which the Parties concerned shall consider in good faith.
9. The expenses of the Commission shall be borne equally by the Parties concerned.
10. When ratifying, accepting, approving or acceding to the present Convention, or at any time thereafter, a Party which is not a regional economic integration organization may declare in a written instrument submitted to the Depositary that, in respect of any dispute not resolved in accordance with paragraph 2, it recognizes as compulsory ipso facto and without special agreement in relation to any Party accepting the same obligation:
    a) Submission of the dispute to the International Court of Justice; and/or
    b) Arbitration by an arbitral tribunal established and operating, unless the Parties to the dispute otherwise agreed, in accordance with the procedure laid down in the annex[14] to this Convention.

    A Party which is a regional economic integration organization may make a declaration with like effect in relation to arbitration in accordance with subparagraph (b).

[14] The annex Arbitration (Article 1-14) has been slightly changed and adapted from: United Nations Environment Programme (UNEP), Convention on Biological Diversity, 05.06.92 (Na.92-7807).

Interim Agreement Between the State of Israel and the Palestinian Authority on the West Bank and the Gaza Strip[15]

signed in Washington, 28 September 1995

Appendix 1
Powers and Responsibilities for Civil Affairs

Article 40
Water and Sewage

On the basis of good-will, both sides have reached the following agreement in the sphere of Water and Sewage:

Principles

1. Israel recognizes the Palestinian water rights in the West Bank. These will be negotiated in the permanent status negotiations and settled in the Permanent Status Agreement relating to the various water resources.
2. Both sides recognize the necessity to develop additional water for various uses.
3. While respecting each side's powers and responsibilities in the sphere of water and sewage in their respective areas, both sides agree to coordinate the management of water and sewage resources and systems in the West Bank during the interim period, in accordance with the following principles:
   a. Maintaining existing quantities of utilization from the resources, taking into consideration the quantities of additional water for the Palestinians from the Eastern Aquifer and other agreed sources in the West Bank as detailed in this Article.
   b. Preventing the deterioration of water quality in water resources.
   c. Using the water resources in a manner which will ensure sustainable use in the future, in quantity and quality.
   d. Adjusting the utilization of the resources according to variable climatological and hydrological conditions.
   e. Taking all necessary measures to prevent any harm to water resources including those utilized by the other side.
   f. Treating, reusing or properly disposing of all domestic, urban, industrial, and agricultural sewage.
   g. Existing water and sewage systems shall be operated, maintained and developed in a coordinated manner, as set out in this Article.

[15] The main text of the Agreement does not address the issue of water resources. The issue was relegated to Appendix 1 [Powers and Responsibilities for Civil Affairs] of Annex III [Protocol concerning Civil Affairs] of the Interim Agreement.

h. Each side shall take all necessary measures to prevent any harm to the water and sewage systems in their respective areas.
i. Each side shall ensure that the provisions of this Article are applied to all resources and systems, including those privately owned or operated, in their respective areas.

Transfer of Authority

4. The Israeli side shall transfer to the Palestinian side, and the Palestinian side shall assume powers and responsibilities in the sphere of water and sewage in the West Bank related solely to Palestinians, that are currently held by the military government and its Civil Administration, except for the issues that will be negotiated in the permanent status negotiations, in accordance with the provisions of this Article.
5. The issue of ownership of water and sewage related infrastructure in the West Bank will be addressed in the permanent status negotiations.

Additional water

6. Both sides agreed that the future water needs of the Palestinians in the West Bank are estimated to be between 70 - 80 mcm/year.
7. In this framework, and in order to meet the immediate needs of the Palestinians in fresh water for domestic use, both sides recognize the necessity to make available to the Palestinians during the interim period a total quantity of 28.6 MCM/year, as detailed below.
   a. Israeli Commitment:
      (1) Additional supply to Hebron and the Bethlehem area, including the construction of the required pipeline - 1 mcm/year.
      (2) Additional supply to Ramallah area - o.5 mcm/year.
      (3) Additional supply to an agreed take-off point in the Salfit area - 0.6 mcm/year.
      (4) Additional supply to the Nablus area - 1 mcm/year.
      (5) The drilling of an additional well in the Jenin area - 1.4 mcm/year.
      (6) Additional supply to the Gaza Strip - 5 mcm/year.
      (7) The capital cost items (1) and (5) above shall be borne by Israel.
   b. Palestinian Responsibility:
      (1) An additional well in the Nablus area - 2.1 mcm/year.
      (2) Additional supply to the Hebron, Bethlehem and Ramallah areas from the Eastern Aquifer or other agreed sources in the West Bank - 17 mcm/year.
      (3) A new pipeline to convey the 5 mcm/year from the existing Israeli water system to the Gaza Strip. In the future, this quantity will come from desalination in Israel.
      (4) The connecting pipeline from the Salfit take-off point to Salfit.
      (5) The connection of the additional well in the Jenin area to the consumers.

(6) The remainder of the estimated quantity of the Palestinians needs mentioned in paragraph 6 above, over the quantities mentioned in this paragraph (41.4 - 51.4 mcm/year), shall be developed by the Palestinians from the Eastern Aquifer and other agreed sources in the West Bank. The Palestinians will have the right to utilize this amount for their needs (domestic and agricultural).

8. The provisions of paragraphs 6-7 above shall not prejudice the provisions of paragraph 1 to this Article.
9. Israel shall assist the Council in the implementation of the provisions of paragraph 7 above, including the following:
   a. Making available all relevant data.
   b. Determining the appropriate occasions for drilling of wells.
10. In order to enable the implementation of paragraph 7 above, both sides shall negotiate and finalize as soon as possible a Protocol concerning the above projects, in accordance with paragraphs 18-19 below.

The Joint Water Committee

11. In order to implement their undertakings under this Article, the two sides will establish, upon the signing of this Agreement, a permanent Joint Water Committee (JWC) for the interim period, under the auspices of the CAC.
12. The function of the JWC shall be to deal with all water and sewage related issues in the West Bank including, inter alia:
    a. Coordinated management of water resources.
    b. Coordinated management of water resources and water and sewage systems.
    c. Protection of water resources and water and sewage systems.
    d. Exchange of information relating to water and sewage laws and regulations.
    e. Overseeing the operation of the joint supervision and enforcement mechanism.
    f. Resolution of water and sewage related disputes.
    g. Cooperation in the field of water and sewage, as detailed in this Article.
    h. Arrangements for water supply from one side to the other.
    i. Monitoring systems. The existing regulations concerning measurement and monitoring shall remain in force until the JWC decides otherwise.
    j. Other issues of mutual interest in the sphere of water and sewage.
13. The JWC shall be comprised of an equal number of representatives from each side.
14. All decisions of the JWC shall be reached by consensus, including the agenda, its procedures and other matters.
15. Detailed responsibilities and obligations of the JWC for the implementation of its functions are set out in Schedule 8.

Supervision and Enforcement Mechanism

16. Both sides recognize the necessity to establish a joint mechanism for supervision over and enforcement of their agreements in the field of water and sewage, in the West Bank.
17. For this purpose, both sides shall establish, upon the signing of this Agreement, Joint Supervision and Enforcement Teams (JSET), whose structure, role, and mode of operation is detailed in Schedule 9.

Water Purchases

18. Both sides agreed that in the case of purchase of water by one side from the other, the purchaser shall pay the full real cost incurred by the supplier, including the cost of production at the source and the conveyance all the way to the point of delivery. Relevant provisions will be included in the Protocol referred to in Paragraph 19 below.
19. The JWC will develop a Protocol relating to all aspects of the supply of water from one side to the other, including, inter alia, reliability of supply, quality of supplied water, schedule of delivery and off-set of debts.

Mutual Cooperation

20. Both sides will cooperate in the field of water and sewage, including, inter alia:
    a. Cooperation in the framework of the Israeli-Palestinian Continuing Committee for Economic Cooperation, in accordance with the provisions of Article XI and Annex III of the Declaration of Principles.
    b. Cooperation concerning regional development programs, in accordance with the provisions of Article XI and Annex IV of the Declaration of Principles.
    c. Cooperation, within the framework of the joint Israeli-Palestinian-American Committee, on water production and development related projects agreed upon by the JWC.
    d. Cooperation in the promotion and development of other agreed water-related and sewage-related joint projects, in existing or future multi-lateral forums.
    e. Cooperation in water-related technology transfer, research and development, training, and setting of standards.
    f. Cooperation in the development of mechanisms for dealing with water-related and sewage-related natural and man-made emergencies and extreme conditions.
    g. Cooperation in the exchange of available relevant water and sewage data, including:
        (1) Measurements and maps related to water resources and uses.
        (2) Reports, plans, studies, researches and project documents related to water and sewage.

(3) Data concerning the existing extractions, utilization and estimated potential of the Eastern, North-Eastern and Western Aquifers (attached as Schedule 10).

Protection of Water Resources and Water and Sewage Systems

21. Each side shall take all necessary measures to prevent any harm, pollution, or deterioration of water quality of the water resources.
22. Each side shall take all necessary measures for the physical protection of the water and sewage systems in their respective areas.
23. Each side shall take all necessary measures to prevent any pollution or contamination of the water and sewage systems, including those of the other side.
24. Each side shall reimburse the other for any unauthorized use of or sabotage to water and sewage systems situated in the areas under its responsibility which serve the other side.

The Gaza Strip

25. The existing agreements and arrangements between the sides concerning water resources and water and sewage systems in the Gaza Strip shall remain unchanged, as detailed in Schedule 11.

Schedule 8
Joint Water Committee

Pursuant to Article 40, paragraph 15 of this Appendix, the obligations and responsibilities of the JWC shall include:

1. Coordinated management of the water resources as detailed hereunder, while maintaining the existing utilization from the aquifers as detailed in Schedule 10, and taking into consideration the quantities of additional water for the Palestinians as detailed in Article 40.

   It is undertstood that the above-mentioned Schedule 10 contains average annual quantities, which shall constitute the basis and guidelines for the operation and decisions of the JWC:

   a. All licensing and drilling of new wells and the increase of extraction from any water source, by either side, shall require the prior approval of the JWC.
   b. All development of water resources and systems, by either side, shall require the prior approval of the JWC.
   c. Notwithstanding the provisions of a. and b. above, it is understood that the projects for additional water detailed in paragraph 7 of Article 40, are agreed in principle between the two sides. Accordingly, only the geohydrological and technical details and specifications of these projects shall be brought before the JWC for approval prior to the commencement of the final design and implementation process.
   d. When conditions, such as climatological or hydrological variability, dictate a reduction or enable an increase in the extraction from a resource, the JWC

shall determine the changes in the extractions and in the resultant supply. These changes will be allocated between the two sides by the JWC in accordance with methods and procedures determined by it.

e. The JWC shall prepare, within three months of the signing of this Agreement, a Schedule to be attached to this Agreement, of extraction quotas from the water resources, based on the existing licenses and permits. The JWC shall update this Schedule on a yearly basis and as otherwise required.

2. Coordinated management of water and sewage systems in the West Bank, as follows:
   a. Existing water and sewage systems, which serve the Palestinian population solely, shall be operated and maintained by the Palestinian side solely, without interference or obstructions, in accordance with the provisions of Article 40.
   b. Existing water and sewage systems serving Israelis, shall continue to be operated and maintained by the Israeli side solely, without interference or obstructions, in accordance with the provisions of Article 40.
   c. The systems referred to in a. and b. above shall be defined on Maps to be agreed upon by the JWC within three months from the signing of this Agreement.
   d. Plans for construction of new water and sewage systems or modification of existing systems require the prior approval of the JWC.

Schedule 9
Supervision and Enforcement Mechanism

Pursuant to Article 40, Paragraph 17 of this Appendix:

1. Both sides shall establish, upon the signing of this Agreement, no less than five Joint Supervision and Enforcement Teams (JSETs) for the West Bank, under the control and supervision of the JWC, which shall commence operation immediately.
2. Each JSET shall be comprised of no less than two representatives form each side, each side in its own vehicle, unless otherwise agreed. The JWC may agree on changes in the number of JSETs and their structure.
3. Each side will pay its own costs, as required to carry out all tasks detailed in this Schedule. Common costs will be shared equally.
4. The JSETs shall operate, in the field, to monitor, supervise and enforce the implementation of Article 40 and this Schedule, and to rectify the situation whenever an infringement has been detected, concerning the following:
   a. Extraction from water resources in accordance with the decisions of the JWC, and the Schedule to be prepared by it in accordance with sub-paragraph 1.e. of Schedule 8.
   b. Unauthorized connections to the supply systems and unauthorized water uses;
   c. Drilling of wells and development of new projects for water supply from all sources;
   d. Prevention of contamination and pollution of water resources and systems;

e. Ensuring the execution of the instructions of the JWC on the operation of monitoring and measurement systems.
f. Operation and maintenance of systems for collection, treatment, disposal and reuse, of domestic and industrial sewage, of urban and agricultural run-off, and of urban and agricultural drainage systems;
g. The electric and energy systems which provide power to all the above systems;
h. The Supervisory Control and Data Acquisition (SCADA) systems for all the above systems;
i. Water and sewage quality analyses carried out in approved laboratories, to ascertain that these laboratories operate according to accepted standards and practices, as agreed by the JWC. A list of the approved laboratories will be developed by the JWC;
j. Any other task, as instructed by the JWC.

5. Activities of the JSETs shall be in accordance with the following:
   a. The JSETs shall be entitled, upon coordination with the relevant DCO, to free, unrestricted and secure access to all water and sewage facilities and systems, including those privately owned or operated, as required for the fulfillment of their function.
   b. All members of the JSET shall be issued identification cards, in Arabic, Hebrew and English containing their full names and a photograph.
   c. Each JSET will operate in accordance with a regular schedule of site visits, to wells, springs and other water sources, water works, and sewage systems, as developed by the JWC.
   d. In addition, either side may require that a JSET visit a particular water or sewage facility or system, in order to ensure that no infringements have occured. When such a requirement has been issued, the JSET shall visit the site in question as soon as possible, and no later than within 24 hours.
   e. Upon arrival at a water or sewage facility or system, the JSET shall collect and record all relevant data, including photographs as required, and ascertain whether an infringement has occured. In such cases, the JSET shall take all necessary measures to rectify it, and reinstate the status quo ante, in accordance with the provisions of this Agreement. If the JSET cannot agree on the actions to be taken, the matter will be referred immediately to the two Chairmen of the JWC for decision.
   f. The JSET shall be assisted by the DCOs and other security mechanisms established under this Agreement, to enable the JSET to implement its functions.
   g. The JSET shall report its findings and operations to the JWC, using forms which will be developed by the JWC.

Schedule 10
Data Concering Aquifers

Pursuant to Article 40, paragraph 20 and Schedule 8 paragraph 1 of this Appendix: The existing extractions, utilization and estimated potential of the Eastern, North-Eastern and Western Aquifers are as follows:

Eastern Aquifer:
- In the Jordan Valley, 40 mcm to Israeli users, from wells;
- 24 mcm to Palestinians, from wells;
- 30 mcm to Palestinians, from springs;
- 78 mcm remaining quantities to be developed from the Eastern Aquifer;

Total: 172 mcm.

North-Eastern Aquifer:
- 103 mcm to Israeli users, from the Gilboa and Beisan springs, including from wells;
- 25 mcm to Palestinian users around Jenin;
- 17 mcm to Palestinian users from East Nablus springs;

Total: 145 mcm.

Western Aquifer:
- 340 mcm used within Israel;
- 20 mcm to Palestinians;
- 2 mcm to Palestinians, from springs near Nablus;

Total: 362 mcm.

All figures are average annual estimates.
The total annual recharge is 679 mcm.

Schedule 11
The Gaza Strip

Pursuant to Article 40, Paragraph 25:
1. All water and sewage (hereinafter referred to as "water") systems and resources in the Gaza Strip shall be operated, managed and developed (including drilling) by the Council, in a manner that shall prevent any harm to the water resources.
2. As an exception to paragraph 1, the existing water systems supplying water to the Settlements and the Military Installation Area, and the water systems and resources inside them shall continue to be operated and managed by Mekoroth Water Co.
3. All pumping from water resources in the Settlements and the Military Installation Area shall be in accordance with the existing quantities of drinking water and agricultural water.

   Without derogating from the powers and responsibilities of the Council, the Council shall not adversely affect these quantities.

   Israel shall provide the Council with all data concerning number of wells in the

Settlements and the quantities and quality of the water pumped from each well, on a monthly basis.

4. Without derogating from the powers and responsibilities of the Council, the Council shall enable the supply of water to the Gush Katif settlement area and Kfar Darom settlement by Mekoroth, as well as the maintenance by Mekoroth of the water systems supplying these locations.
5. The council shall pay Mekoroth for the cost of water suplied from Israel and for the real expenses incurred in supplying water for the Council.
6. All relations between the Council and Mekoroth shall be dealt with in a commercial agreement.
7. The Council shall take the necessary measures to ensure the protection of all water systems in the Gaza Strip.
8. The two sides shall establish a subcommittee to deal with all issues of mutual interest including the exchange of all relevant data to the management and operation of the water resources and systems and the mutual prevention of harm to water resources.
9. The subcommittee shall agree upon its agenda and upon the prodecures and manner of its meetings, and may invite experts or advisers as it sees fit.

Agreement Between the State of Israel and the Palestinian Authority on the Gaza Strip and the Jericho Area

signed in Cairo, 4 May 1994

[...] 31. Water and Sewage

a. All water and sewage (hereinafter referred to as 'water') systems and resources in the Gaza Strip and the Jericho Area shall be operated, managed and developed (including drilling) by the Palestinian Authority, in a manner that shall prevent any harm to the water resources.
b. As an exception to subparagraph a., the existing water systems supplying water to the Settlements and the Military Installation Area, and the water systems and resources inside them continue to be operated and managed by Mekoroth Water Co.
c. All pumping from water resources in the Settlements and the Military Installation Area, shall be in accordance with existing quantities of drinking water and agricultural water. Without derogating from powers and responsibilities of the Palestinian Authority, the Palestinian Authority shall not adversely affect these quantities. Israel shall provide the Palestinian Authority with data concerning the number of wells in the Settlements and the quantities of the water pumped from each well, on a monthly basis.
d. Without derogating from powers and responsibilities of the Palestinian Authority, the Palestinian Authority shall enable the supply of water to the Gush Katif settlement area and the Kfar Darom settlement by Mekoroth, as well as the maintenance by Mekoroth of the water systems supplying these locations and of water lines crossing the Jericho Area.
e. The Palestinian Authority shall pay Mekoroth for the cost of water supplied from Israel and for the real expenses incurred in supplying water to the Palestinian Authority.
f. All relations between the Palestinian Authority and Mekoroth shall be dealt with in a commercial agreement.
g. The Palestinian Authority shall take the necessary measures to ensure the protection of all water systems in the Gaza Strip and the Jericho Area.
h. Upon the signing of this Agreement, the two Parties shall establish a subcommittee to deal with all issues of mutual interest including the exchange of all data relevant to the management and operation of the water resources and systems and mutual prevention of harm to water resources.
i. The subcommittee shall agree upon its agenda and upon the procedures and manner of its meetings, and may invite experts or advisers as it sees fit.

Treaty of Peace Between the State of Israel and the Hashemite Kingdom of Jordan (26 October 1994)

Annex II
Water-related Matters

Pursuant to Article 6 of the treaty, Jordan and Israel agreed on the following articles on water-related matters:

Article I
Allocation

1. Water from the Yarmouk River
   a. Summer period - 15th May to 15th October of each year. Israel pumps (12) mcm and Jordan gets the rest of the flow.
   b. Winter period - 16th October to 14th May of each year. Israel pumps (13) mcm and Jordan is entitled to the rest of flow subject to provisions outlined hereinbelow: Jordan concedes to Israel pumping an additional (20) mcm from the Yarmouk in winter in return for Israel conceding to transferring to Jordan during the summer period the quantity specified in paragraph (2,A) below from the Jordan River.
   c. In order that waste of water will be minimised, Jordan and Israel may use, downstream of Adassiya diversion/point 121, excess flood water that is not usable and will evidently go to waste unused.
2. Water from the Jordan River
   a. Summer period - 15th May to 15th October of each year. In return for the additional water that Jordan concedes to Israel in winter in accordance with paragraph (1,B) above, Israel concedes to transfer to Jordan in the summer period (20) mcm form the Jordan River directly upstream from Deganya gates on the river. Jordan shall pay the operation and maintenance cost of such transfer through existing systems (not including capital cost) and shall bear the total cost of any new transmission system. A separate protocol shall regulate this transfer.
   b. Winter period - 16th October to 14th May of each year. Jordan is entitled to store for its use a minimum average of (20) mcm of the floods in the Jordan River south of its confluence with the Yarmouk (as outlined in Article II). Excess floods that are not usable and that will otherwise be wasted can be utilised for the benefit of the two parties including pumping storage off the course of the river.
   c. In addition to the above, Israel is entitled to maintain its current uses of the Jordan River waters between its confluence with the Yarmouk and its confluence with Wadi Yabis/Tirat Zvi. Jordan is entitled to an annual quantity equivalent to that of Israel, provided however, that Jordan's use will not harm the quantity or quality of the above Israeli uses. The joint water

committee (outlined in Article VI) will survey existing uses for documentation and prevention of appreciable harm.

d. Jordan is entitled to an annual quantity of (9) mcm of desalinated water from the desalination of about (20) mcm of saline springs now diverted to the Jordan River. Israel will explore the possibility of financing the operation and maintenane cost of the supply to Jordan of this desalinated water (not including capital cost). Until the desalination facilities are operational, and upon the entry into force of the treaty, Israel will supply Jordan (9) mcm of Jordan River water from the same location as in (2,A) above, outside the summer period and during dates Jordan selects, subject to the maximum capacity of transmission.

3. Additional water

Jordan and Israel will cooperate in finding sources for the supply to Jordan of an additional quantity of 50 mcm/yr of water to drinkable standards. To this end, the joint water committee will develop, within one year from the entry into force of the treaty, a plan for the supply to Jordan of the above mentioned additional water. This plan will be forwarded to the respective governments for discussion and decision.

4. Operation and Maintenance
   a. Operation and maintenance of the systems on Israeli territory that supply Jordan with water, and their electricity supply, shall be Israel's responsibility. The operation and maintenance of the new systems, that serve only Jordan will be contracted at Jordan's expense to authorities or companies selected by Jordan.
   b. Israel will guarantee easy unhindered access of personnel and equipment to such new systems for operation and maintenance. This subject will be further detailed in the agreements to be signed between Israel and the authorities or companies selected by Jordan.

Article II
Storage

1. Jordan and Israel shall cooperate to build a diversion/storage dam on the Yarmouk River directly downstream of the Adassiya diversion/point 121. The purpose is to improve the diversion efficiency into the King Abdullah Canal of the water allocation of the Hashemite Kingdom of Jordan, and, possibly, for the diversion of Israel's allocation of the river water. Other purposes can be mutually agreed.
2. Jordan and Israel shall cooperate to build a system of water storage on the Jordan River, along their common boundary, between its confluence with the Yarmouk River and its confluence with Wadi Yabis/Tirat Zvi, in order to implement the provision of paragraph (2,B) of Article 1 above. The storage system can also be made to accomodate more floods; Israel may use up to 3 mcm/year of added storage capacity.
3. Other storage reservoirs can be discussed and agreed upon mutually.

Article III
Water Quality and Protection

1. Jordan and Israel each undertake to protect, within their own jurisdiction, the shared waters of the Jordan and Yarmouk Rivers, and Araba/Arava groundwater, against any pollution, contamination, harm or unauthorised withdrawals of each other allocations.
2. For this purpose, Jordan and Israel will jointly monitor the quality of water along their boundary, by use of jointly established monitoring stations to be operated under the guidance of the Joint Water Committee.
3. Jordan and Israel will each prohibit the disposal of municipal and industrial wastewater into the course of the Yarmouk or the Jordan rivers before they are treated to standards allowing their unrestricted agricultural use. Implementation of this prohibition shall be completed within three years from the entry into force of the Treaty.
4. The quality of water supplied from one country to the other at any given location shall be equivalent to the quality of the water used from the same location by the supplying country.
5. Saline springs currently diverted to the Jordan River are earmarked for desalination within four years. Both countries shall cooperate to ensure that the resulting brine (?) will not be disposed of in the Jordan River or in any of its tributaries.
6. Jordan and Israel will each protect water systems in its own territory supplying water to the other, against any pollution, contamination, harm or unauthorised withdrawal of each other allocations.

Article IV
Groundwater in Wadi Araba/Emek Ha'arava

1. In accordance with the provision of this treaty, some wells drilled and used by Israel along with their associated systems fall on the Jordanian side of the borders. These wells and systems are under Jordan's sovereignty. Israel shall retain the use of these wells and systems in the quantity and quality detailed in Appendix 1, that shall be jointly prepared by December 31, 1994. Neither country shall take, nor cause to be taken, any measure that may appreciably reduce the yields or quality of these wells and systems.
2. Throughout the period of Israel's use of these wells and systems, replacement of any well that may fail among them shall be licensed by Jordan in accordance with the laws and regulations then in effect. For this purpose, the failed well shall be treated as though it was drilled under license from the competent Jordanian authority at the time of its drilling. Israel shall supply Jordan with the log of each of the wells and the technical information about it to be kept on record. The replacement well shall be connected to the Israeli electricity and water systems.
3. Israel may increase the abstraction rate from wells and systems in Jordan by up to 9 mcm/yr above the yields referred to in paragraph 1 above, subject to a determination by the joint water committee that this undertaking is hydrogeologi-

cally feasible and does not harm existing Jordanian uses. Such increase is to be carried out within five years from the entry into force of the treaty.

4. Operation and maintenance
   a. Operation and maintenance of the wells and systems on Jordanian territory that supply Israel with water, and their electricity supply shall be Jordan's responsibility. The operation and maintenance of these wells and systems will be contracted at Israel's expense to authorities or companies selected by Israel.
   b. Jordan will guarantee easy unhindered access of personnel and equipment to such wells and systems for operation and maintenance. This subject will be further detailed in the agreements to be signed between Jordan and the authorities or companies selected by Israel.

Article V
Notification and Agreement

1. Artificial changes in or of the course of the Jordan and Yarmouk rivers can only be made by mutual agreement.
2. Each country undertakes to notify the other, six months ahead of time, of any intended projects which are likely to change the flow of either of the above rivers along their commin boundary, or the quality of such flow. The subject will be discussed in the joint water committee with the aim of preventing harm and mitigating adverse impacts such projects may cause.

Article VI
Cooperation

1. Jordan and Israel undertake to exchange relevant data on water resources through the joint water committee.
2. Jordan and Israel shall cooperate in developing plans for purpose of increasing water supplies and improving water use efficiency, within the context of bilateral, regional or international cooperation.

Article VII
Joint Water Committee

1. or the purpose of the implementation of this annex, the parties establish a joint water committee comprised of three members from each country.
2. The joint water committee will, with the approval of the respective governments, specify its work procedures, the frequency of its meetings, and the details of its scope of work. The committee may invite experts and/or advisors as may required.
3. The committee may form, as it seems necessary, a number of specialised subcommittees and assign them technical tasks. In this context, it is agreed that these subcommittees will include a northern subcommittee and a southern subcommittee, for the management on the ground of the mutual water resources in these sectors.

Agreement Between the Hashemite Kingdom of Jordan and the Syrian Republic for the Utilization of the Yarmouk River[17]

Signed in Amman, 3 September 1987

The Government of the Hashemite Kingdom of Jordan and the Government of the Syrian Republic;

Confirming bonds of friendship and ties of Arab race existing between the two countries, and emhasizing the spirit of sincere cooperation which exists between them, and after perusal of the results of discussions which took place between their two delegations in Damascus on July 4-5, 1987 and August 9-11, 1987 for the utilization of the waters of the Yarmouk basin; and in accordance with the agreement signed in Damascus June 4, 1953;

And appreciating the benefits which could be secured for the two countries by the proper collection and utilization of the waters of the Yarmmouk basin for the purpose of securing irrigation of agricultural lands and the generation of electrical power;

Have resolved to conclude an Agreement and have delegated their representatives for this purpose:

For the Government of the Hashemite Kingdom of Jordan:
Zaid Al-Rifai, Prime Minister

For the Government of the Syrian Republic:
Dr. Abdul Raouf Al-Kassem, Prime Minister

And after having examined the credentials which each holds on behalf of his Government and after being satisfied with the genuineness of these documents, these representatives have agreed on the following articles:

Article 1

In this treaty the following terms shall have the meanings as defined hereunder:

a. "Jordan" - The Government of the Hashemite Kingdom of Jordan.
b. "Syria" - The Government of the Syrian Republic.
c. "Government" - Jordan or Syria as the context requires.
d. "Jordan Valley" - The Valley of the River Jordan.
e. "Al-Wihda (Unity) Dam and Reservoir" - A dam on the Yarmouk River for collection of water, and a reservoir for water storage to be constructed on Jordanian and Syrian Territory.
f. "Al-Wihda Power Plant" - A plant for the generation of electricity located on the Southern bank of the Yarmouk River within the Al-Wihda Dam.

[17] English translation from the Arabic original.

g. "Yarmouk Project" - Al-Wihda dam and Reservoir, the Al-Wihda Power Plant, and any other buildings and construction required by this project and the relocation of the Hejaz Railway near Maqarin Station.
h. "The Joint Commission" - The Jordanian-Syrian Commission defined in Article 9 of this Agreement.

Article 2

The two governments recognize that, for natural and technical reasons, it is important to obtain the additional water and the electrical power which both countries need in an economical and practical way, by construction of the Al-Wihda dam. For this purpose they agree to the construction of the following works:

a. The Al-Wihda Dam and Reservoir which are a dam designed to collect river water and a reservoir to be constructed on the Yarmouk River on Jordanian and Syrian territory near Maqarin Station in Syria. This water is intended for the generation of electrical power and for use in irrigating Jordanian lands and for other Jordanian schemes and for irrigating Syrian lands located downstream from the dam along the river bed up to a level of 200 meter above sea level.
b. The Hydroelectric Power Plant which is to be constructed below the dam to generate whatever electric power is possible.
c. Relocation of the Hejaz Railway where it follows the Yarmouk Valley as necessitated by the project and the construction of all buildings and works required by the project.

Article 3

Jordan agrees, subject to the provisions of Article 9 of this Agreement, to assume responsibility for the cost of investigations, studies, design construction, operation and maintenance.

Syria agrees to provide all necessary facilities and assistance in allowing project workers to enter Syrian territory and to carry out thereon all work relating to the project through all stages of the work, subject to the regulations in force in its Government and in accordance with this Agreement.

Article 4

Each government will compensate the owners of the lands and properties expropriated for the Yarmouk Project in accordance with the laws in effect by both Governments. Syria will settle all claims related to water rights and compensate the owners, and Jordan will bear the responsibility to pay all compensations and expenditures paid by Syria in relation to acquisition and other rights.

Article 5

Jordanian and Syrian laborers will be employed in the construction of the Yarmouk Project as necessary. Jordanian and Syrian technicians shall be emloyed during the period of studies, execution and maintenance. Syrian public construction companies will be treated equal to Jordanian construction companies at the construction of this project.

Article 6

Jordan will carry out design and construction of the Al-Wihda Dam at a height of 100 meter including a gated spillway for the storage of water of the Yarmouk River, after filling with water all Syrian dams listed in the attached annex. Syria has the right to hold all stored water in these dams as an integral part of their right in this agreement. Design of Al-Wihda Dam should provide for the future raising of the Dam to store more water if it is found feasible, technical and economical by both Governments.

Article 7

a. Syria reserves the right to dispose of the waters of all streams springing in its lands in the basin of the Yarmouk and its tributaries, with the exception of waters having their source upstream of the dam below the 250 meter contour. Syria also reserves the right to utilize the waters of the River or from tributaries which join the River below the dam for irrigating Syrian lands situated in the lower basin of the River.
b. Jordan has the right to dispose of the water released from the Reservoir and the Power Plant at the Al-Wihda Dam for the purpose of generating electrical power.
c. Electricity generated from the Al-Wihda Dam will be divided between Jordan and Syria in the ratio of 25 percent to Jordan and 75 percent to Syria.

Article 8

Syria will be responsible for the implementation of the relocation of the Hejaz Railway as necessitated by the project and the construction of all railways, buildings and work required by the project, and Jordan will pay for all costs required to implement it.

Article 9

A Jordanian-Syrian commission, consisting of citizens of both Governments, shall be formed for the execution of the provisions of this Agreement, the regulation of rights and obligations which the two Governments have acquired and accepted therefrom, and for exercising these rights and obligations and for considering all cases which may arise from their application.

This Joint Commission shall be considered a legal body (a body possessing juristic personality), the members of which shall enjoy diplomatic privileges and rights in the Government which they do not represent. The Joint Commission shall consist of three members from each Government, with the provision that one of them shall be of the grade of Deputy-Minister, or Under-Secretary. The Joint Commission shall have the right to seek advice of experts and advisers, and to employ assistants, technicians, and employees from among the citizens of both Governments and other Governments as required for carrying out its work.

Representation of this commission to outsiders shall be by both presidents jointly and not separately.

The Joint Commission shall be the responsible body in accordance with the provisions of this agreement, and shall carry out all responsibilities entrusted to it for the completion of the project and the utilization of its benefits in the best manner possible. In the event of the inability of the Joint Commission to reach a definite agreement, the members shall immediately submit a report to this effect to their respective Governments and the two Governments shall then settle the dispute immediately and find the subjective solutions which guarantee the proper performance of the work and insure the rights of both parties according to the rules of the agreement by direct contact of both parties.

The joint commission shall set up internal regulations for organizing its work. These regulations shall be approved by the heads of both Governments.

Article 10

Employees, representatives of both Governments, members of the Joint Commission, and employees of technical organizations working on the project, who are in possession of an authorization from the Joint Commission to that effect, and who carry special identification documents, shall have the right to visit the lands on which the Al-Wihda project works are located, together with such neighboring lands as the Joint Commission may find necessary for the purpose of carrying out studies, investigations, construction work, administration, and maintenance, without being subject to procedures arising from the application of the passports or any other law or regulation in force by the two Governments provided that local laws of the two Governments shall remain in force in their own territories in all other respects.

Article 11

The two Governments agree that, each within its territory, shall take all necessary measures for the prevention or minimization of silt accumulation in the Interstate Reservoir, and shall also take effective measures for erosion control, forestation, grass development, gully control and other measures which will assist in attaining the maximum benefit from the full capacity of the reservoir. The cost of these measures shall be paid by Jordan.

Article 12

The rights of the two Governments in the use of the lake in its territory, to utilize,, operate and maintain for the purpose of tourism and raising fish in a way which does not interfere with the management of the Al-Wihda Dam facilities.

Article 13

Boundaries between the two Governments shall remain unchanged as before the construction of the Al-Wihda Dam and shall be considered drawn on the surface of the water.

Article 14

This Agreement shall be ratified by the competent authorities of the two Governments and shall become valid from the date of exchange of ratification documents. Its provisions, which may be amended by annexes, shall continue in force until the treaty is nullified by another concluded between the two Governments for this purpose.

In confirmation of what is stated above, the representatives of the two Governments have signed and sealed this Agreement with their respective signatures and seals.

This Agreement is drawn up in Arabic with two original copies which are equally ruling.

Article 15

The Agreement signed between the two Governments in Damascus on June 4, 1953 for the utilization of the waters of the Yarmouk River shall be herewith cancelled.

Signed in Amman on the 3rd of September 1987

For the Government of the Hashemite Kingdom of Jordan
Zaid Rifai, Prime Minister

For the Government of the Syrian Republic
Dr. Abdul Raouf Al-Kassem, Prime Minister

Annex

The Annex lists the names, height above sea level and storage capacities of 25 dams located on Syrian territory on the following northern tributaries of the Yarmouk River: Al-Thabah [5 dams], Al-Zeidie [5 dams], Al-Aram [4 dams], Al-Allan [4 dams] and Al-Ruqad [7 dams] rivers. The total capacity of these dams is 61.7 MCM, of which the Kudneh dam on the Al-Ruqad River alone accounts for 30 MCM.

Agreement Between the Republic of the Sudan and the United Arab Republic for the Full Utilization of the Nile Waters[17]

Signed in Cairo, 8 November 1959

As the River Nile needs projects, for its full control and for increasing its yield for the full utilization of its waters by the Republic of the Sudan and the United Arab Republic on technical working arrangements other than those applied:

And as these works require their execution and administration, full agreement and cooperation between the two Republics in order to regulate their benefits and utilize the Nile waters in a manner which secures the present and future requirements of the two countries:

And as the Nile waters Agreement concluded in 1929[18] provided only for the partial use of the Nile waters and did not include a complete control of the River waters, the two Republics have agreed on the following:

First
The Present Acquired Rights

1. That the amount of the Nile waters used by the United Arab Republic until this Agreement is signed shall be her acquired right before obtaining the benefits of the Nile Control Projects and the projects which will increase its yield and which projects referred to in this Agreement; the total of this acquired right is 48 billion of cubic meters per year as measured at Aswan.
2. That the amount of the waters used at present by the Republic of Sudan shall be her acquired right before obtaining the benefits of the projects referred to above. The total amount of this acquired right is 4 billion of cubic meters per year as measured at Aswan.

Second
The Nile Control Projects and the Division of Their Benefits Between the two Republics

1. In order to regulate the River waters and control their flow into the sea, the two Republics agree that the United Arab Republic constructs the Sudd el Aali at Aswan as the first link of a serie of projects on the Nile for over-year storage.
2. In order to enable the Sudan to utilize its share of the water, the two Republics agree that the Republic of Sudan shall construct the Roseires Dam on the Blue Nile and any other works which the Republic of the Sudan considers essential for the utilization of its share.
3. The net benefit from the Sudd el Aali Reservoir shall be calculated on the basis of the average natural River yield of water at Aswan in the years of this cen-

[17] United Nations Treaty Series 453, No.6519, 1963, pp.64-74. Official text: Arabic, registered by the United Arab Republic on 7 February 1963.

[18] Came into force on 12 December 1959 in accordance with Article 7.

tury, which is estimated at about 84 billion of cubic meters per year. The acquired rights of the two Republics referred to in Article 'First' as measured at Aswan, and the average losses of over-year storage of the Sudd el Aali Reservoir shall be deducted from this yield, and the balance shall be the net benefit which shall be divided between the two Republics.

4. The net benefit from the Sudd el Aali Reservoir mentioned in the previous item, shall be divided between the two Republics at the ratio of 14.5 for the Sudan and 7.5 for the United Arab Republic so long as the average river yield remains in future within the limits of the average yield referred to in the previous paragraph. This means that, if the average yield remains the same as the average of the previous years of this century which is estimated at 84 billion, and if the losses of over-year storage remain equal to the present estimate of 10 billion, the net benefit of the Sudd el Aali Reservoir shall be 22 billion of which the share of the Republic of the Sudan shall be 14.5 billion and the share of the United Arab Republic shall be 7.5 billion. By adding these shares to their acquired rights, the total share from the net yield of the Nile after the full operation of the Sudd el Aali Reservoir shall be 18.5 billion for the Republic of Sudan and 55.5 billion for the United Arab Republic. But if the average yield increases, the resulting net benefit from this increase shall be divided between the two Republics, in equal shares.
5. As the net benefit from the Sudd el Aali (referred to in item 3 Article Second) is calculated on the basis of the average natural yield of the river at Aswan in the years of this century after the deduction therefrom of the acquired rights of the two Republics and the average losses of over-year storage at the Sudd el Aali Reservoir, [...] at reasonable intervals to be agreed upon after starting the full operation of the Sudd el Aali Reservoir.
6. The United Arab Republic agrees to pay to the Sudan Republic 15 million Egyptian Pounds as full compensation for the damage resulting to the Sudanese existing properties as a result of the storage in the Sudd el Aali Reservoir up to a reduced level of 182 metres (survey datum). The payment of this compensation shall be affected in accordance with the annexed agreement between the two parties.
7. The Republic of the Sudan undertakes to arrange before July 1963, the final transfer of the population of Halfa and all other Sudanese inhabitants whose lands shall be submerged by the stored water.
8. It is understood that when the Sudd el Aali is fully operated for over-year storage, the United Arab Republic will not require storing any water at Gebel Aulia Dam. And the two contracting parties will in due course, discuss all matters related to this renunciation.

Third
Projects for the Utilization of Lost Waters in the Nile Basin

In view of the fact that at present, considerable volumes of the Nile Basin Waters are lost in the swamps of Bahr el Jebel, Bahr el Zeraf, Bahr el Ghazal and the Sobat River, and as it is essential that efforts should be exerted in order to prevent these losses and to increase the yield of the River for use in agricultural expansion in the two Republics, the two Republics agree to the following:

1. The Republic of the Sudan in agreement with the United Arab Republic shall construct projects for the increase of the River yield by preventing losses of waters of the Nile Basin in the swamps of Bahr el Jebel, Bahr el Zeraf, Bahr el Ghazal and its tributaries, the Sobat River and its tributaries and the White Nile Basin. The net yield of these projects shall be divided equally between the two Republics and each of them shall also contribute equally to the costs. The Republic of the Sudan shall finance the above-mentioned projects out of its own funds and the United Arab Republic shall pay its share in the costs in the same ratio of 50% allotted for her in the yield of these projects.
2. If the United Arab Republic, on account of the progress in its planned agricultural expansion should find it necessary to start on any increase of the Nile yield projects, referred to in the previous paragraph, after its approval by the two Governments and at a time when the Sudan Republic does not need such project, the United Arab Republic shall notify the Sudan Republic of the time convenient for the former to start the execution of the project. And each of the two Republics shall, within two years after such notification, present a date-phased programme for the utilization of its share of the waters saved by the project, and each of the said programmes shall bind the two parties. The United Arab Republic shall at the expiry of the two years, start the execution of the projects, at its own expense. And when the Republic of Sudan is ready to utilize its share according to the agreed programme, it shall pay to the United Arab Republic a share of all the expenses in the same ratio as the Sudan's share in benefit is to the total benefit of the project; provided that the share of either Republic shall not exceed one half of the total benefit of the project.

Fourth
Technical Cooperation Between the two Republics

1. In order to ensure the technical cooperation between the Governments of the two Republics, to continue the research and study necessary for the Nile control projects and the increase of its yield and to continue hydrological survey of its upper reaches, the two Republics agree that immediately after signing of this Agreement a Permanent Joint Technical Commission shall be formed of an equal number of members from both parties; and its functions shall be:
   a) The drawing of the basic outlines of projects for the increase of the Nile yield, and for the supervision of the studies necessary for the finalising of projects, before presentation of the same to the Governments of the two Republics for approval.
   b) The supervision of the execution of the projects approved by the Governments.
   c) The drawing up of the working arrangements for any works to be constructed on the Nile, within the boundaries of the Sudan, and also for those to be constructed outside the boundaries of the Sudan, by agreement with

the authorities concerned in the countries in which such works are constructed.

d) The supervision of the application of all the working arrangements mentioned in (c) above in connection with works constructed within the boundaries of Sudan and also in connection with the Sudd el Aali Reservoir and Aswan Dam, through official engineers delegated for the purpose by the two Republics; and the supervision of the working of the upper Nile projects, as provided in the agreements concluded with the countries in which such projects are constructed.

e) As it is probable that a series of low years my occur, and a succession of low levels in the Sudd el Aali Reservoir may result to such an extent as not to permit in any one year the drawing of the full requirements of the two Republics, the Technical Commission is charged with the task of devising a fair arrangement for the two Republics to follow. And the recommendations of the Commission shall be presented to the two Governments for approval.

2. In order to enable the Commission to exercise the functions enumerated in the above item, and in order to ensure the continuation of the Nile gauging and to keep observations on all its upper reaches, these duties shall be carried out under the technical supervision of the Commission by the engineers of the Sudan Republic, and the engineers of the United Arab Republic in the Sudan and in the United Arab Republic and in Uganda.
3. The two Governments shall form the Joint Technical Commission, by a joint decree, and shall provide it with its necessary funds from their budgets. The Commission may, according to the requirements of work, hold its meetings in Cairo or in Khartoum. The Commission shall, subject to the approval of the two Governments, lay down regulations for the organisation of its meetings and its technical, administrative and financial activities.

Fifth
General Provisions

1. If it becomes necessary to hold any negotiations concerning the Nile waters, with any riparian state, outside the boundaries of the two Republics, the Governments of the Sudan Republic and the United Arab Republic shall agree on a unified view after the subject is studied by the said Technical Commission. The said unified view shall be the basis of any negotiations by the Commission with the said states.
   If the negotiations result in an agreement to construct any works on the river, outside the boundaries of the two Republics, the Joint Technical Commission shall after consulting the authorities in the Governments of the States concerned, draw all the technical execution details and the working and maintenance arrangements. And the Commission shall, after the sanction of the same by the Governments concerned, supervise the carrying out of the said technical agreements.
2. As the riparian states, other than the two Republics, claim a share in the Nile waters, the two Republics have agreed that they shall jointly consider and reach

one unified view regarding the said claims. And if the said consideration results in the acceptance of allotting an amount of the Nile water to one or the other of the said states, the accepted amount shall be deducted from the shares of the two Republics in equal parts, as calculated at Aswan.

The Technical Commission mentioned in this agreement shall make the necessary arrangments with the states concerned, in order to ensure that their water consumption shall not exceed the amounts agreed upon.

Sixth
Transitional Period Before Benefiting From the Complete Sudd El Aali Reservoir

As the benefiting of the two Republics from their appointed shares in the net benefit of the Sudd el Aali Reservoir shall not start before the construction and the full utilization of the Reservoir, the two parties shall agree on their agricultural expansion programmes in the transitional period from now up to the completion of the Sudd el Aali, without prejudice to their present water requirements.

Seventh

This Agreement shall come into force after its sanction by the two contracting parties, provided that either party shall notify the other party of the date of its santion, through diplomatic channels.

Eighth

Annex (1) and Annex (2, A and B) attached to this Agreement shall be considered as an integral part of this Agreement.

Written in Cairo in two Arabic original copies this 7th day of Gumada El Oula 1379, the 8th day of November 1959.
For the Republic of Sudan: (Signed) Lewa Mohammed Talaat Farid
For the United Arab Republic: (Signed) Zakaria Mohie El Din

Printing: Mercedesdruck, Berlin
Binding: Buchbinderei Lüderitz & Bauer, Berlin